T0281421

Rasch Measurement Theory Analysis in R

Chapman & Hall/CRC
The R Series

Series Editors
John M. Chambers, Department of Statistics, Stanford University, California, USA
Torsten Hothorn, Division of Biostatistics, University of Zurich, Switzerland
Duncan Temple Lang, Department of Statistics, University of California, Davis, USA
Hadley Wickham, RStudio, Boston, Massachusetts, USA

Recently Published Titles

For more information about this series, please visit: https://www.crcpress.com/Chapman--Hall-CRC-The-R-Series/book-series/CRCTHERSER

Rasch Measurement Theory Analysis in R

Stefanie A. Wind and Cheng Hua

CRC Press
Taylor & Francis Group
Boca Raton London New York

CRC Press is an imprint of the
Taylor & Francis Group, an **informa** business

A CHAPMAN & HALL BOOK

First edition published 2022
by CRC Press
6000 Broken Sound Parkway NW, Suite 300, Boca Raton, FL 33487-2742

and by CRC Press
4 Park Square, Milton Park, Abingdon, Oxon, OX14 4RN

CRC Press is an imprint of Taylor & Francis Group, LLC

Library of Congress Cataloging-in-Publication Data

Names: Wind, Stefanie A. (Anne), author. | Hua, Cheng, author.
Title: Rasch measurement theory analysis in R / Stefanie Wind and Cheng Hua.
Description: First edition. | Boca Raton : CRC Press, [2022] | Series: Chapman and Hall/CRC the R series | Includes bibliographical references and index.
Identifiers: LCCN 2021058803 (print) | LCCN 2021058804 (ebook) | ISBN 9780367776398 (pbk) | ISBN 9781032005607 (hbk) | ISBN 9781003174660 (ebk)
Subjects: LCSH: Psychometrics. | Rasch models. | Psychology--Statistical methods. | Social sciences--Statistical methods.
Classification: LCC BF39 .W564 2022 (print) | LCC BF39 (ebook) | DDC 150.1/5195--dc23/eng/20220303
LC record available at https://lccn.loc.gov/2021058803
LC ebook record available at https://lccn.loc.gov/2021058804

ISBN: 978-1-032-00560-7 (hbk)
ISBN: 978-0-367-77639-8 (pbk)
ISBN: 978-1-003-17466-0 (ebk)

DOI: 10.1201/9781003174660

Typeset in Latin Modern
by KnowledgeWorks Global Ltd.

Publisher's note: This book has been prepared from camera-ready copy provided by the authors.

Contents

1

Introduction

The purpose of this book is to illustrate techniques for conducting Rasch measurement theory (Rasch, 1960) analyses using existing R packages. The book includes some background information about Rasch models, and the primary objectives are to demonstrate how to apply the models to data using R packages and how to interpret the results.

The primary audience for this book is graduate students or professionals who are familiar with Rasch measurement theory at a basic level, and who want to use the R software program (R Core Team, 2021) to conduct their Rasch analyses. We provide a brief overview of several key features of Rasch measurement theory in this chapter, and we provide descriptions of basic characteristics of the models and analytic techniques in each of the following chapters. Accordingly, we encourage readers who are new to Rasch measurement theory to use this book as a supplement to other excellent introductory texts on the subject that include a detailed theoretical and statistical introduction to Rasch measurement. For example, interested readers may find the following texts useful to begin learning about Rasch measurement theory:

Andrich, David, and Ida Marais. 2019. *A Course in Rasch Measurement Theory: Measuring in the Educational, Social and Health Sciences.* Singapore: Springer.

Bond, Trevor G., Zi Yan, and Moritz Heene. 2019. *Applying the Rasch Model: Fundamental Measurement in the Human Sciences (4th Ed.).* New York: Routledge, Taylor & Francis Group.

Engelhard, George, and Jue Wang. *Rasch Models for solving measurement problems: Invariant Measurement in the Social Sciences.* Vol.187. SAGE, 2020. https://us.sagepub.com/en-us/nam/rasch-models-for-solving-measurement-problems/book267292

This book also assumes a basic working knowledge of the R software and programming language. To use this book, readers will need to know how to run existing code in R or R Studio, and how to make basic edits to existing code in order to adapt it for use with their own data. Readers who are new to R may find the following resources helpful for learning how to use this program:

- What is R & R Studio: https://libguides.library.kent.edu/statconsulting/r

- Install R & R-Studio: https://rstudio-education.github.io/hopr/starting. html

- R Tutorial for beginners: https://rstudio-education.github.io/hopr/starting. html

In addition, readers should note that our descriptions of the R code generally assume that the analyses are being conducted in R Studio. However, all of the R code will work in both R and R Studio.

1.1 Overview of Rasch Measurement Theory

Georg Rasch was a Danish psychometrician who introduced a theory and approach to social science measurement in his classic text entitled *Probabilistic Models for Some Intelligence and Attainment Tests* (Rasch, 1960). This approach to measurement involves transforming ordinal item responses, such as the data that are collected in a multiple-choice educational assessment of middle school students' understanding of engineering design (Alemdar et al., 2017), a survey designed to measure self-efficacy for making career decisions (Nam et al., 2011), or a diagnostic scale used to identify individuals with depression (Shea et al., 2009), to interval-level measures for examinees and items. Now called Rasch measurement theory, this approach is based on principles and requirements that reflect measurement in the physical sciences.

Chief among the defining features of Rasch measurement theory is the emphasis on *invariance* in measurement. In the context of Rasch measurement theory, invariance occurs when the following properties are observed in item response data (Rasch, 1961):

- The comparison between two stimuli should be independent of which particular individuals were instrumental for the comparison;

- and it should also be independent of which stimuli within the considered class were or might also have been compared.

- Symmetrically, a comparison between two individuals should be independent of which particular stimuli with the class considered were instrumental for the comparison;

- and it should also be independent of which other individuals were also compared on the same or on some other occasion (pp. 331-332)

Rasch used the term *specific objectivity* (Rasch, 1977) to describe the importance of identifying specific situations in which the requirements for invariant measurement are approximated. In emphasizing invariance, Rasch noted that meaningful interpretation and use of social science measurement instruments is not possible unless invariance is approximated.

Rather than assuming that data will adhere perfectly to the model requirements, researchers who use Rasch models do so in order to identify deviations from these requirements when they occur. Information about departures from model requirements can help analysts identify areas for additional research, including qualitative investigations of persons and items, as well as guidance for improving the quality of a measurement procedure. This perspective, in which the *measurement theory* (i.e., the model) is emphasized as a guide for understanding the quality of the data by comparing it to strict requirements, is a key distinguishing feature between Rasch measurement theory and other item response theory (IRT) approaches. In typical IRT analyses, analysts attempt to identify a model that is the most accurate representation of the characteristics of the *data* (Embretson and Reise, 2000). For example, many researchers select the three-parameter logistic model (Birnbaum, 1968) when analyzing responses to multiple-choice educational assessments because the model directly incorporates instances of guessing and differences in item discrimination. However, researchers guided by a Rasch perspective would instead use the Rasch model to identify unexpected observations that could alert them to potential guessing and inconsistent item ordering over examinee achievement levels (as reflected by differences in item discrimination). These unexpected observations could then lead to additional exploration and the improvement of the assessment procedure. Although there are many situations in which reproducing the characteristics of item response data may be useful or necessary, the general perspective that characterizes Rasch measurement theory is that the theory (as reflected in the model) provides a framework for evaluating data according to its adherence to fundamental measurement properties. Rasch measurement theory scholars argue that evidence of adherence to model requirements is necessary before data can be used to make inferences about persons and items (e.g., in statistical analyses). As Bond and Fox (2019) noted, "researchers should spend more time investigating their scales than investigating with their scales" (p. 4).

In addition to providing useful information about adherence to fundamental measurement properties, Rasch models have several other theoretical and practical features that have contributed to their widespread popularity across disciplines in the social, behavioral, and health sciences (please see Rasch, 1977 for a review). Wright and Mok (2004) summarized the key theoretical and practical features of the Rasch measurement approach as follows:

In order to construct inference from observation, the measurement model must: (a) produce linear measures, (b) overcome missing data, (c) give estimates of precision, (d) have devices for detecting misfit, and (e) the parameters of the object being measured and of the measurement instrument must be separable. Only the family of Rasch measurement models solves these problems (p. 4).

To help researchers take advantage of these useful features in a practical way, our book provides an overview of several key models within the family of Rasch models, offers basic guidance on the estimation of the models using available R packages, and provides suggestions and advice for interpreting the results from the analyses.

1.2 Online Resources

This book includes several supplemental resources that are available online, including copies of the R code, example data sets, and data sets for the challenge exercises at the end of some of the chapters. All of the materials used in this book, including the R software, R packages, and data sets, are free to download. Table 1.1 provides details about how to download all the relevant learning materials.

TABLE 1.1
Online Resources

Title	Version	Download Link
R Programming Language	4.0.3 (latest)	cran.r-project.org
R Studio	1.3	rstudio.com
Rasch Book Online Version	Beta 0.2	https://beta.rstudioconnect.com/ connect/#/apps/e5fb8a2a-e6d7- 4ede-8e53-888cf0129c9e/access
Source Code for this Book	GitHub	https://github.com/ huacheng1985/ Bookdown_CRC_Rasch
Data set	Beta	See in Each Chapter

https://www.routledge.com/Rasch-Measurement-Theory-Analysis-in-R/ Wind-Hua/p/book/9781032005607

2

Dichotomous Rasch Model

This chapter provides a basic overview of the dichotomous Rasch model, along with guidance for analyzing data with the dichotomous Rasch model using R. We use data from a transitive reasoning assessment presented by Sijtsma and Molenaar (2002) to illustrate the analysis using Conditional Maximum Likelihood Estimation (CMLE) with the Extended Rasch Modeling (eRm) package (Mair et al., 2021). Then, we illustrate the application of the dichotomous Rasch model using Marginal Maximum Likelihood Estimation (MMLE) and Joint Maximum Likelihood Estimation (JMLE) with the Test Analysis Modules (TAM) package (Robitzsch et al., 2021). After the analyses are complete, we present an example description of the results. The chapter concludes with a challenge exercise and resources for further study.

Overview of the Dichotomous Rasch Model

The *dichotomous Rasch model* (Rasch, 1960) is the simplest model in the Rasch family of models (Wright and Mok, 2004). It was designed for use with ordinal data that are scored in two categories (usually 0 or 1). The dichotomous Rasch model uses sum scores from these ordinal responses to calculate interval-level estimates that represent person locations (i.e., person ability or person achievement) and item locations (i.e., the difficulty to provide a correct or positive response) on a linear scale that represents the latent variable (the log-odds or "logit" scale). The difference between person and item locations can be used to calculate the probability for a correct or positive response ($x = 1$), rather than an incorrect or negative response ($x = 0$).

The equation for the dichotomous Rasch model can be expressed in log-odds form as follows:

$$\ln\left[\frac{\phi_{ni1}}{\phi_{ni0}}\right] = \theta_n - \delta_i \tag{2.1}$$

The Rasch model predicts the probability that person n on item i provides a correct or positive ($x = 1$) response, rather than an incorrect or negative ($x = 0$) response, given person locations (i.e., ability, achievement, θ_n) and item locations (i.e., difficulty, δ_i), as expressed on the logit scale.

DOI: 10.1201/9781003174660-2

Rasch Model Requirements

Estimates that are calculated using the dichotomous Rasch model can only be meaningfully interpreted if there is evidence that the data approximate the requirements for the model. Key among dichotomous Rasch model requirements are the following:

- *Unidimensionality*: A single latent variable is sufficient to explain most of the variation in item responses.
- *Local independence*: After controlling for the latent variable, there is no substantial association between the responses to individual items.
- *Person-invariant item estimates*: Item locations do not depend on (i.e., are independent from) the persons whose responses are used to estimate them.
- *Item-invariant person estimates*: Person locations do not depend on (i.e., are independent from) the items used to estimate them.

Evidence that data approximate these requirements provides support for the meaningful interpretation and use of item and person estimates on the logit scale as indicators of item and person locations on the latent variable. In practice, many analysts evaluate some or all of these requirements using various indicators of model-data fit for the facets in a Rasch model (in this case, items and persons). In the current chapter, we provide some basic code for calculating some popular residual-based fit indices for items and persons. We explore issues related to model requirements and evaluating model-data fit in more detail in Chapter 3.

2.1   Example Data: Transitive Reasoning Test

In this chapter, we will be working with data from a transitive reasoning test designed to measure students' ability to reason about the relationships among physical objects. The data were collected from a one-on-one interactive assessment in which an experimenter presented students with a set of objects, such as sticks, balls, cubes, and discs. The following description is given in Sijtsma and Molenaar (2002), pp. 31–32:

> The items for transitive reasoning had the following structure. A typical item used three sticks, here denoted A, B, and C, of different length, denoted Y, such that YA < YB < YC. The actual test taking had the form of a conversation between experimenter and child in which the sticks were identified by their colors rather than letters. First, sticks A and B were presented to a child, who was allowed to pick them up and compare their lengths, for example, by placing them next to each other on a table.

Next, sticks B and C were presented and compared. Then all three sticks were displayed in a random order at large mutual distances so that their length differences were imperceptible, and the child was asked to infer the relation between sticks A and C from his or her knowledge of the relationship in the other two pairs.

The transitive reasoning items varied in terms of the property students were asked to reason about (length, weight, area). The tasks also varied in terms of the number of physical objects that students were asked to reason about, and whether the comparison tasks involved equalities, inequalities, or a mixture of equalities and inequalities. The characteristics of the transitive reasoning data are summarized in the following table:

Task	Property	Format	Objects	Measures
1	Length	YA > YB > YC	Sticks	12, 11.5, 11 (cm)
2	Length	YA = YB = YC = YD	Tubes	12 (cm)
3	Weight	YA > YB > YC	Tubes	45, 25, 18 (g)
4	Weight	YA = YB = YC = YD	Cubes	65 (g)
5	Weight	YA < YB < YC	Balls	40, 50, 70 (g)
6	Area	YA > YB> YC	Discs	2.5, 7, 6.5 (diameter; cm)
7	Length	YA > YB = YC	Sticks	28.5, 27.5, 27.5 (cm)
8	Weight	YA >YB = YC	Balls	65, 40, 40 (g)
9	Length	YA = YB = YC = YD	Sticks	12.5, 12.5, 13, 13 (cm)
10	Weight	YA = YB < YC = YD	Balls	60, 60, 100, 100 (g)

2.2 Dichotomous Rasch Model Analysis with CMLE in eRm

In the next section, we provide a step-by-step demonstration of a dichotomous Rasch model analysis using the *eRm* package (Mair et al., 2021), which uses CMLE. We encourage readers to use the example data set that is provided in the online supplement to conduct the analyses along with us.

Prepare for the Analyses

We selected eRm for the first illustration in the current chapter because it includes functions for applying the dichotomous Rasch model that are relatively straightforward to use and interpret. Please note that the eRm package uses the CMLE method to estimate Rasch model parameters. As a result, estimates from the eRm package are not directly comparable to estimates obtained using other estimation methods. Later in this chapter, we have included illustrations of dichotomous Rasch model analyses with the TAM package (Robitzsch et al., 2021) with MMLE. We also provide an illustration with TAM using JMLE, which produces comparable estimates to JMLE estimates from some popular standalone Rasch software programs, such as Winsteps (Winsteps, 2020b) and Facets (Facets, 2020a).

To get started with the eRm package, install and load it into your R environment using the following code.

```
#install.packages("eRm")
library("eRm")
```

Now that we have installed and loaded the package to our R session, we are ready to import the data.

In this book, we use the function `read.csv()` to import data that are stored using comma separated values. We encourage readers to use their preferred method for importing data files into R or R Studio. Please note that if you use `read.csv()` you will need to first specify the file path to the location at which the data file is stored on your computer or set your working directory to the folder in which you have saved the data.

First, we will import the data using `read.csv()` and store it in an object called `transreas`.

```
transreas <- read.csv("transreas.csv")
```

Next, we will explore the data using descriptive statistics using the `summary()` function.

```
summary(transreas)
```

```
##      Student              Grade              task_01
##   Min.    :  1     Min.    :2.000     Min.    :0.0000
##   1st Qu.:107     1st Qu.:3.000     1st Qu.:1.0000
##   Median :213     Median :4.000     Median :1.0000
##   Mean    :213     Mean    :4.005     Mean    :0.9412
##   3rd Qu.:319     3rd Qu.:5.000     3rd Qu.:1.0000
##   Max.    :425     Max.    :6.000     Max.    :1.0000
##      task_02              task_03
##   Min.    :0.0000     Min.    :0.0000
##   1st Qu.:1.0000     1st Qu.:1.0000
##   Median :1.0000     Median :1.0000
##   Mean    :0.8094     Mean    :0.8847
##   3rd Qu.:1.0000     3rd Qu.:1.0000
##   Max.    :1.0000     Max.    :1.0000
##      task_04              task_05
```

```
##    Min.    :0.0000    Min.    :0.0000
##    1st Qu.:1.0000    1st Qu.:1.0000
##    Median :1.0000    Median :1.0000
##    Mean    :0.7835    Mean    :0.8024
##    3rd Qu.:1.0000    3rd Qu.:1.0000
##    Max.    :1.0000    Max.    :1.0000
##       task_06          task_07
##    Min.    :0.0000    Min.    :0.0000
##    1st Qu.:1.0000    1st Qu.:1.0000
##    Median :1.0000    Median :1.0000
##    Mean    :0.9741    Mean    :0.8447
##    3rd Qu.:1.0000    3rd Qu.:1.0000
##    Max.    :1.0000    Max.    :1.0000
##       task_08          task_09          task_10
##    Min.    :0.0000    Min.    :0.0000    Min.    :0.00
##    1st Qu.:1.0000    1st Qu.:0.0000    1st Qu.:0.00
##    Median :1.0000    Median :0.0000    Median :1.00
##    Mean    :0.9671    Mean    :0.3012    Mean    :0.52
##    3rd Qu.:1.0000    3rd Qu.:1.0000    3rd Qu.:1.00
##    Max.    :1.0000    Max.    :1.0000    Max.    :1.00
```

From the summary of **transreas**, we can see there are no missing data. We can also get a general sense of the scales, range, and distribution of each variable in the data set.

Specifically, we can see that Student ID numbers range from 1 to 425, student grade levels range from 2 to 6, and all tasks have scores in both of the dichotomous categories (0 and 1). We can also get a sense for the range of item difficulty by examining the mean for each task, which is the proportion-correct statistic (item difficulty estimate for Classical Test Theory).

Run the Dichotomous Rasch Model

To run the dichotomous Rasch Model using the eRm package, need to isolate the item response matrix from the other variables in the data (student IDs and grade level). To do this, we will create an object made up of only the item responses by removing the first two variables from the data. We will remove the descriptive variables using the **subset()** function with the **select=** option. We will save the response matrix in a new object called **transreas.responses**.

```
transreas.responses <- subset(transreas,select=-c(
    Student,Grade))
```

Next, we will use **summary()** to calculate descriptive statistics for the **transreas.responses** object to check our work and ensure that the responses are ready for analysis.

```
summary(transreas.responses)
```

```
##       task_01                  task_02
##   Min.    :0.0000    Min.    :0.0000
##   1st Qu.:1.0000    1st Qu.:1.0000
##   Median :1.0000    Median :1.0000
##   Mean    :0.9412    Mean    :0.8094
##   3rd Qu.:1.0000    3rd Qu.:1.0000
##   Max.    :1.0000    Max.    :1.0000
##       task_03                  task_04
##   Min.    :0.0000    Min.    :0.0000
##   1st Qu.:1.0000    1st Qu.:1.0000
##   Median :1.0000    Median :1.0000
##   Mean    :0.8847    Mean    :0.7835
##   3rd Qu.:1.0000    3rd Qu.:1.0000
##   Max.    :1.0000    Max.    :1.0000
##       task_05                  task_06
##   Min.    :0.0000    Min.    :0.0000
##   1st Qu.:1.0000    1st Qu.:1.0000
##   Median :1.0000    Median :1.0000
##   Mean    :0.8024    Mean    :0.9741
##   3rd Qu.:1.0000    3rd Qu.:1.0000
##   Max.    :1.0000    Max.    :1.0000
##       task_07                  task_08
##   Min.    :0.0000    Min.    :0.0000
##   1st Qu.:1.0000    1st Qu.:1.0000
##   Median :1.0000    Median :1.0000
##   Mean    :0.8447    Mean    :0.9671
##   3rd Qu.:1.0000    3rd Qu.:1.0000
##   Max.    :1.0000    Max.    :1.0000
##       task_09                  task_10
##   Min.    :0.0000    Min.    :0.00
##   1st Qu.:0.0000    1st Qu.:0.00
##   Median :0.0000    Median :1.00
##   Mean    :0.3012    Mean    :0.52
##   3rd Qu.:1.0000    3rd Qu.:1.00
##   Max.    :1.0000    Max.    :1.00
```

Now, we are ready to run the dichotomous Rasch model on the transitive reasoning response data. We will use the RM() function to run the model and store the results in an object called dichot.transreas.

```
dichot.transreas <- RM(transreas.responses)
```

Overall Model Summary

The next step is to request a summary of the model estimation results in order to begin to understand the results from the analysis. We can do so by applying the summary() function to the model object.

```
summary(dichot.transreas)
```

```
##
## Results of RM estimation:
##
## Call:  RM(X = transreas.responses)
##
## Conditional log-likelihood: -921.3465
## Number of iterations: 18
## Number of parameters: 9
##
## Item (Category) Difficulty Parameters (eta): with
##   0.95 CI:
##          Estimate Std. Error lower CI upper CI
## task_02     0.258      0.133   -0.003    0.518
## task_03    -0.416      0.157   -0.723   -0.109
## task_04     0.441      0.128    0.190    0.692
## task_05     0.309      0.131    0.052    0.567
## task_06    -2.175      0.292   -2.747   -1.604
## task_07    -0.025      0.141   -0.302    0.252
## task_08    -1.909      0.262   -2.423   -1.395
## task_09     2.923      0.130    2.668    3.179
## task_10     1.836      0.115    1.610    2.062
##
## Item Easiness Parameters (beta) with 0.95 CI:
##                  Estimate Std. Error lower CI
## beta task_01        1.243      0.204    0.842
## beta task_02       -0.258      0.133   -0.518
## beta task_03        0.416      0.157    0.109
## beta task_04       -0.441      0.128   -0.692
## beta task_05       -0.309      0.131   -0.567
## beta task_06        2.175      0.292    1.604
## beta task_07        0.025      0.141   -0.252
## beta task_08        1.909      0.262    1.395
## beta task_09       -2.923      0.130   -3.179
## beta task_10       -1.836      0.115   -2.062
##                  upper CI
## beta task_01        1.643
## beta task_02        0.003
## beta task_03        0.723
```

```
## beta task_04    -0.190
## beta task_05    -0.052
## beta task_06     2.747
## beta task_07     0.302
## beta task_08     2.423
## beta task_09    -2.668
## beta task_10    -1.610
```

The summary of the dichotomous Rasch model output includes the Conditional Log-likelihood statistic, details about the number of iterations and model parameters, and a table with item parameters, their standard errors, and confidence intervals. It is important to note that the item parameters included in this preliminary output are *item easiness* parameters– *not* item difficulty parameters. We will examine item difficulty parameters in detail later in our analysis.

Wright Map

A useful feature of Rasch models is that when there is acceptable fit between the model and the data (discussed in detail in Chapter 3), it is possible to visualize and compare item and person locations on a single linear continuum. Professor Benjamin D. Wright popularized an approach to displaying Rasch model results on a linear continuum, and Rasch measurement researchers across disciplines have adopted this technique. In his honor, many researchers refer to these displays as *Wright Maps*. Researchers also refer to these displays as *Variable Maps*. Please see Wilson (2011) for a discussion of the term *Wright Map*.

As the next step in our analysis, we will create a Wright Map from our model results. We will create the plot using the function `PlotPImap()` on the model object (`dichot.transreas`). We will set the option for displaying threshold labels as `FALSE`, because we are working with dichotomous data. We also used the `main.title=`option to customize the title of the plot.

```
plotPImap(dichot.transreas, main = "Transitive
    Reasoning Assessment Wright Map")
```

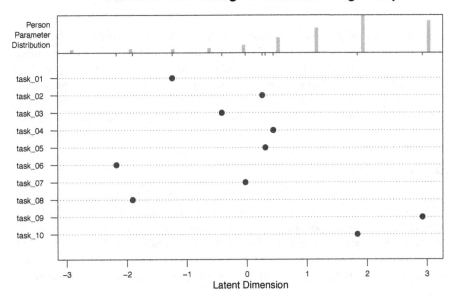

In this *Wright Map* display, the results from the dichotomous Rasch model analysis of the Transitive Reasoning data are summarized graphically. The figure is organized as follows:

Starting at the bottom of the figure, the horizontal axis (labeled *Latent Dimension*) is the logit scale that represents the latent variable. In the application of the Transitive Reasoning data, lower numbers indicate less transitive reasoning ability, and higher numbers indicate more transitive reasoning ability.

The central panel of the figure shows item difficulty locations on the logit scale for the 10 transitive reasoning tasks that were included in the analysis; the y-axis for this panel shows the item labels. By default in eRm, the items are ordered according to their original order in the response matrix. The items can be ordered by difficulty by adding `sorted = TRUE` as an argument in the `plotPImap()` call. For each item, a solid circle plotting symbol shows the location estimate.

The upper panel of the figure shows a histogram of person (in this case, student) location estimates on the logit scale. Small vertical lines on the x-axis of this histogram show the points on the logit scale at which information (variance) is maximized for the sample of persons and items in the analysis. These lines can be omitted by adding `irug = FALSE` as an argument in the `plotPImap()` call.

Even though it is not appropriate to fully interpret item and person locations on the logit scale until there is evidence of acceptable model-data fit, we recommend examining the Wright Map during the preliminary stages of an item analysis to get a general sense of the model results and to identify any potential scoring or data entry errors.

A quick glance at the Wright Map suggests that, on average, the persons (students) are located higher on the logit scale compared to the average item (task) locations. In addition, there appears to be a relatively wide spread of person and item locations on the logit scale, such that the transitive reasoning test appears to be a useful tool for identifying differences in person locations and item locations on the latent variable. We will return to this display for more exploration after we check for acceptable model-data fit, among other psychometric properties.

Item Parameters

As the next step in our analysis, we will examine the item parameters in detail. The eRm package provides several options with which analysts can find and examine item location parameters. For example, one way to obtain the overall item location parameters is to extract the *eta parameters* from the model object using the $ operator. We extract these parameters, print them to the console, and calculate summary statistics for them with the following code.

```
difficulty <- dichot.transreas$etapar
difficulty
```

```
##       task_02       task_03       task_04       task_05
##    0.25775001   -0.41581384   0.44093780   0.30941151
##       task_06       task_07       task_08       task_09
##   -2.17531698   -0.02481129  -1.90884851   2.92343872
##       task_10
##    1.83581566
```

Because of the nature of the estimation process used in eRm, the item.locations object that we just created does not include the location estimate for the first item. One can calculate the location for item 1 by subtracting the sum of the item locations from zero. In the following code, we find the location for item 1, and then create a new object with all 10 item locations.

```
n.items <- ncol(transreas.responses)
i1 <- 0 - sum(difficulty[1:(n.items - 1)])
difficulty.all <- c(i1, difficulty[c(1:(n.items - 1))
    ])
difficulty.all
```

```
##                     task_02       task_03       task_04
##   -1.24256308    0.25775001   -0.41581384   0.44093780
##       task_05       task_06       task_07       task_08
##    0.30941151   -2.17531698   -0.02481129  -1.90884851
##       task_09       task_10
##    2.92343872    1.83581566
```

Alternatively, we could find the item difficulty parameters by extracting the item *easiness* parameters from the model object (*beta parameters*) and multiplying them by −1 to find item difficulty.

```
difficulty2 <- dichot.transreas$betapar * -1
difficulty2
```

```
## beta task_01 beta task_02 beta task_03
##  -1.24256308    0.25775001   -0.41581384
## beta task_04 beta task_05 beta task_06
##   0.44093780    0.30941151   -2.17531698
## beta task_07 beta task_08 beta task_09
##  -0.02481129   -1.90884851    2.92343872
## beta task_10
##   1.83581566
```

The item difficulty parameters (xsi) are the item location estimates on the logit scale that represents the latent variable. Assuming that the responses are scored such that lower scores ($x = 0$) indicate lower locations on the latent variable (e.g., incorrect, negative, or absent responses), *higher* item estimates on the logit scale indicate items that are *more difficult* or require persons to have relatively *higher locations* on the construct to provide a correct response. On the other hand, *lower* item estimates on the logit scale indicate items that are *easier* or require persons to have relatively *lower locations* on the construct to provide a correct response. In our analysis, Task 9 is the most difficult item ($\delta = 2.92$), whereas Task 6 is the easiest item ($\delta = -2.18$).

Next, we will examine item standard errors. The standard errors are reported for all of the items as part of the beta (item easiness) parameters.

```
dichot.transreas$se.beta
```

```
## [1] 0.2043488 0.1327967 0.1566695 0.1282289
## [5] 0.1314316 0.2916385 0.1413876 0.2622114
## [9] 0.1302767 0.1152173
```

The standard error for each item is an estimate of the precision of the item estimates, where larger standard errors indicate less-precise estimates. Standard errors are reported on the same logit scale as item locations. In our analysis, the standard errors range from 0.11 for Task 9 and Task 10, which were the items with the most precise estimates, to 0.31 for Task 6, which was the item with the least precise estimate. These differences largely reflect differences in item targeting to the person locations (see the Wright Map above).

Next, let's calculate descriptive statistics to better understand the distribution of the item locations and standard errors. We will do so using the summary() function and the sd() function.

```
summary(difficulty.all)
```

```
##     Min. 1st Qu.  Median    Mean 3rd Qu.    Max.
## -2.1753 -1.0359  0.1165  0.0000  0.4081  2.9234
```

```
sd(difficulty.all)
```

```
## [1] 1.576441
```

```
summary(dichot.transreas$se.beta)
```

```
##     Min. 1st Qu.  Median    Mean 3rd Qu.    Max.
##   0.1152  0.1306  0.1371  0.1694  0.1924  0.2916
```

```
sd(dichot.transreas$se.beta)
```

```
## [1] 0.06206381
```

We can also visualize the item difficulty estimates using a simple histogram by using the hist() function. In our plot, we specified a custom title using main = and a custom x-axis label using xlab =:

```
hist(difficulty.all, main = "Histogram of Item
    Difficulty Estimates for the \nTransitive
    Reasoning Data",
      xlab = "Item Difficulty Estimates Logits")
```

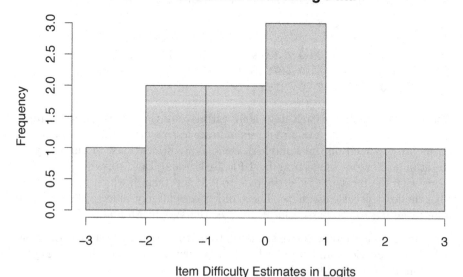

Histogram of Item Difficulty Estimates for the Transitive Reasoning Data

Item Difficulty Estimates in Logits

Item Response Functions

We will examine graphical displays of item difficulty using item response functions (IRFs). With the dichotomous Rasch model, the eRm package creates plots of the probability for a correct or positive response ($x = 1$), conditional on person locations on the latent variable. In the following code, we use `plotICC()` from eRm to create IRF plots for the items in our analysis. We included `ask = FALSE` in our function call to generate all of the plots at once. For brevity, we have only included plots for the first three items in this book. The specific items to be plotted can be controlled by changing the items included in the `items.to.plot` object.

```
items.to.plot <- c(1:3)
plotICC(dichot.transreas, ask = FALSE, item.subset =
    items.to.plot)
```

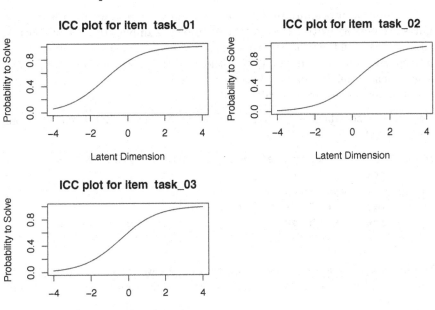

This code generates IRFs for the first three items in our analysis. In each item-specific plot, the x-axis is the logit scale that represents the latent variable; this scale represents transitive reasoning ability in our example. The y-axis is the probability for a correct or positive response, conditional on person locations on the latent variable.

Person Parameters and Item Fit

In the eRm package, it is necessary to calculate person parameters *before* item fit statistics can be calculated. Accordingly, we will proceed with a brief examination of person parameters before we conduct item fit analyses. In practice, we recommend examining item fit before examining and interpreting item nad person locations in detail.

Person Parameters

In the following code, we use the `person.parameter()` function to estimate student locations and save the results in an object called `student.locations`. Then, we save the theta estimates (i.e., achievement estimates) in a data.frame object called `achievement`, and add student identification numbers for reference.

```
student.locations <- person.parameter(dichot.
    transreas)
achievement <- student.locations$theta.table
achievement$id <- rownames(achievement)
```

The estimation procedure in eRm does not directly produce parameter estimates for persons with extreme scores. In our example, 51 students earned extreme scores on the transitive reasoning assessment, so achievement estimates are reported for the remaining 374 students with non-extreme scores. The achievement estimates for the 51 students with extreme scores are extrapolated, and standard errors are not calculated.

We can add standard errors for the student achievement estimates to our `achievement` object as follows.

```
se <- as.data.frame(student.locations$se.theta$
    NAgroup1)
se$id <- rownames(se)
names(se) <- c("person_se", "id")
achievement.with.se <- merge(achievement, se, by = "
    id" )
```

As a final step in examining the person parameters, we will calculate descriptive statistics and use a histogram to examine the distribution of student locations on the logit scale.

```
summary(achievement)
```

```
##   Person Parameter      NAgroup     Interpolated
##   Min.    :-2.916   Min.    :1   Mode  :logical
##   1st Qu.: 1.160    1st Qu.:1   FALSE:374
##   Median : 1.932    Median :1   TRUE  :51
##   Mean    : 2.010   Mean    :1
##   3rd Qu.: 3.028    3rd Qu.:1
##   Max.    : 4.201   Max.    :1
##         id
##   Length:425
##   Class  :character
```

```
##   Mode    : character
##
##
##
```

```
hist(achievement$`Person Parameter`, main = "
    Histogram of Person Achievement Estimates \nfor
    the Transitive Reasoning Data",
    xlab = "Person Achievement Estimates Logits")
```

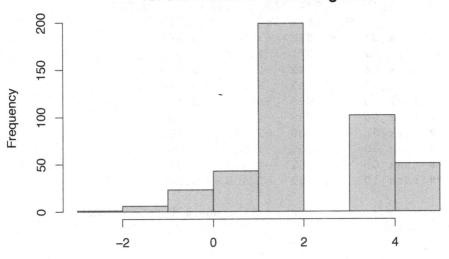

Histogram of Person Achievement Estimates for the Transitive Reasoning Data

Person Achievement Estimates in Logits

Item Fit

Next, we will conduct a brief exploration of item fit statistics. We explore item fit in more detail in Chapter 3.

To calculate numeric item fit statistics, we will use the function `itemfit()` from eRm on the person parameter object (`person.locations.estimate`). This function produces several item fit statistics, including infit mean square error (MSE), outfit MSE, and standardized infit and outfit MSE statistics. We will store the item fit results in a new object called `item.fit`, and then format this object as a data.frame object for easy manipulation and exporting.

```
student.locations <- person.parameter(dichot.
    transreas)
item.fit <- itemfit(student.locations)
item.fit
```

```
##
## Itemfit Statistics:
##              Chisq  df p-value Outfit MSQ Infit MSQ
## task_01 245.550 373   1.000      0.657     0.761
## task_02 421.904 373   0.041      1.128     1.087
## task_03 218.155 373   1.000      0.583     0.745
## task_04 478.894 373   0.000      1.280     1.158
## task_05 346.015 373   0.839      0.925     0.991
## task_06  65.613 373   1.000      0.175     0.611
## task_07 244.494 373   1.000      0.654     0.804
## task_08 125.478 373   1.000      0.336     0.687
## task_09 394.207 373   0.216      1.054     0.904
## task_10 326.534 373   0.960      0.873     0.898
##           Outfit t Infit t Discrim
## task_01    -1.327  -1.622   0.455
## task_02     1.071   1.187   0.054
## task_03    -2.742  -2.688   0.583
## task_04     2.442   2.268  -0.014
## task_05    -0.625  -0.109   0.184
## task_06    -2.789  -1.781   0.620
## task_07    -2.768  -2.475   0.493
## task_08    -2.205  -1.569   0.563
## task_09     0.462  -1.727   0.022
## task_10    -1.822  -2.408   0.226
```

Next, we will request a summary of the numeric fit statistics using the summary () function.

```
summary(item.fit$i.infitMSQ)
```

```
##    Min. 1st Qu.  Median    Mean 3rd Qu.    Max.
##  0.6114  0.7490  0.8506  0.8645  0.9689  1.1575
```

```
summary(item.fit$i.infitZ)
```

```
##     Min. 1st Qu.  Median     Mean 3rd Qu.    Max.
## -2.6885 -2.2509 -1.6743 -1.0924 -0.4742  2.2680
```

```
summary(item.fit$i.outfitMSQ)
```

```
##    Min. 1st Qu.  Median    Mean 3rd Qu.    Max.
##  0.1754  0.6009  0.7648  0.7665  1.0218  1.2805
```

```
summary(item.fit$i.outfitZ)
```

```
##     Min. 1st Qu.  Median     Mean 3rd Qu.    Max.
## -2.7885 -2.6080 -1.5748 -1.0304  0.1904  2.4418
```

The `item.fit` object includes mean square error (MSE) and standardized (Z) versions of the outfit and infit statistics for each item included in the analysis. These statistics are summaries of the residuals associated with each item. When data fit Rasch model expectations, the MSE versions of outfit and infit are expected to be close to 1.00 and the standardized versions of outfit and infit are expected to be around 0.00. Please refer to Chapter 3 for a more-detailed discussion of item fit.

As a final step in examining item fit, we will calculate the *reliability of separation statistic for items* using the following procedure.

```
# Get Item scores
ItemScores <- colSums(transreas.responses)

# Get Item Standard Deviation (SD)
ItemSD <- apply(transreas.responses,2,sd)

# Calculate the Standard Error (SE) for the Items
ItemSE <- ItemSD/sqrt(length(ItemSD))

# Calculate the Observed Variance (also known as
    Total Person Variability or Squared Standard
    Deviation)
SSD.ItemScores <- var(ItemScores)

# Calculate the Mean Square Measurement error (also
    known as Model Error variance)
Item.MSE <- sum((ItemSE)^2) / length(ItemSE)

# Calculate the Item Separation Reliability
item.separation.reliability <- (SSD.ItemScores-Item.
    MSE) / SSD.ItemScores
item.separation.reliability
```

```
## [1] 0.9999984
```

Briefly, the *item reliability of separation statistic* describes the degree to which items have unique locations on the logit-scale. Please refer to Chapter 3 for more details about item reliability analysis.

Person Fit

Next, we will conduct a brief exploration of person fit. To calculate numeric person fit statistics, we will use the function `personfit()` from eRm on the person parameter object (`person.locations.estimate`). This function produces several person fit statistics, including infit MSE, outfit MSE, and standardized infit and outfit MSE statistics. We will store the person fit results in a new object called `person.fit`, and then format this object as a data.frame for easy manipulation and exporting. Then, we request summaries of the infit MSE and outfit MSE statistics using `summary()`.

```
person.fit <- personfit(student.locations)
summary(person.fit$p.infitMSQ)
```

```
##     Min. 1st Qu.  Median    Mean 3rd Qu.     Max.
##   0.4060  0.4910  0.7487  0.9102  1.1740  2.2483
```

```
summary(person.fit$p.outfitMSQ)
```

```
##     Min. 1st Qu.  Median    Mean 3rd Qu.     Max.
##   0.1723  0.2399  0.5468  0.7665  0.9592  7.3060
```

We can calculate the *reliability of person separation statistic* using eRm. This value is interpreted similarly to Cronbach's alpha (Cronbach, 1951) when there is good fit between the data and the Rasch model (Andrich, 1982). However, it is important to note that these coefficients are not equivalent because alpha is based on an assumption of linearity and the Rasch reliability of separation statistic is based on a linear, interval-level scale when there is evidence of good model-data fit (discussed in Chapter 3).

```
person_rel <- SepRel(student.locations)
person_rel$sep.rel
```

```
## [1] 0.2061478
```

Summarize the Results in Tables

As a final step, we will create tables that summarize the calibrations of the items and persons from our dichotomous Rasch model analysis with eRm.

Table 1 is an overall model summary table that provides an overview of the logit-scale locations, standard errors, fit statistics, and reliability statistics for items and persons. This type of table is useful for reporting the results from Rasch model analyses because it provides a concise overview of the location estimates and numeric model-data fit statistics for the items and persons in the analysis.

```
summary.table.statistics <-
  c("Logit Scale Location Mean",
  "Logit Scale Location SD",
  "Standard Error Mean",
  "Standard Error SD",
  "Outfit MSE Mean",
  "Outfit MSE SD",
  "Infit MSE Mean",
```

```
    "Infit MSE SD",
    "Std. Outfit Mean",
    "Std. Outfit SD",
    "Std. Infit Mean",
    "Std. Infit SD",
    "Reliability of Separation")

item.summary.results <-
  rbind(mean(difficulty.all),
        sd(difficulty.all),
        mean(dichot.transreas$se.beta),
        sd(dichot.transreas$se.beta),
        mean(item.fit$i.outfitMSQ),
        sd(item.fit$i.outfitMSQ),
        mean(item.fit$i.infitMSQ),
        sd(item.fit$i.infitMSQ),
        mean(item.fit$i.outfitZ),
        sd(item.fit$i.outfitZ),
        mean(item.fit$i.infitMSQ),
        sd(item.fit$i.infitZ),
        item.separation.reliability)


person.summary.results <-
  rbind(mean(achievement.with.se$`Person Parameter`),
        sd(achievement.with.se$`Person Parameter`),
        mean(achievement.with.se$person_se, na.rm =
            TRUE),
        sd(achievement.with.se$person_se, na.rm =
            TRUE),
        mean(person.fit$p.outfitMSQ),
        sd(person.fit$p.outfitMSQ),
        mean(person.fit$p.infitMSQ),
        sd(person.fit$p.infitMSQ),
        mean(person.fit$p.outfitZ),
        sd(person.fit$p.outfitZ),
        mean(person.fit$p.infitZ),
        sd(person.fit$p.infitZ),
        person_rel$sep.rel)

# Round the values for presentation in a table:
item.summary.results_rounded <- round(item.summary.
    results, digits = 2)
```

```
person.summary.results_rounded <- round(person.
    summary.results, digits = 2)

Table1 <- cbind.data.frame(summary.table.statistics,
                           item.summary.results_
                           rounded,
                           person.summary.results_
                           rounded)

# Add descriptive column labels:
names(Table1) <- c("Statistic", "Items", "Persons")
print.data.frame(Table1, row.names = FALSE)

##                          Statistic Items Persons
##      Logit Scale Location Mean      0.00    1.71
##        Logit Scale Location SD      1.58    1.09
##             Standard Error Mean     0.17    0.95
##              Standard Error SD      0.06    0.16
##               Outfit MSE Mean       0.77    0.77
##                Outfit MSE SD        0.35    0.80
##                Infit MSE Mean       0.86    0.91
##                 Infit MSE SD        0.18    0.47
##             Std. Outfit Mean       -1.03    0.08
##               Std. Outfit SD        1.83    0.64
##               Std. Infit Mean       0.86   -0.12
##                Std. Infit SD        1.67    0.91
##      Reliability of Separation      1.00    0.21
```

Table 2 summarizes the calibrations of individual items. For data sets with manageable item sample sizes such as the transitive reasoning data example in this chapter, we recommend reporting details about each item in a table similar to this one.

```
# Calculate the proportion correct for each task:
PropCorrect <- apply(transreas.responses, 2, mean)

# Combine item calibration results in a table:
Table2 <-
  cbind.data.frame(colnames(transreas.responses),
                   PropCorrect,
                   difficulty.all,
                   dichot.transreas$se.beta,
                   item.fit$i.outfitMSQ,
                   item.fit$i.outfitZ,
                   item.fit$i.infitMSQ,
                   item.fit$i.infitZ)
```

```
names(Table2) <- c("Task ID", "Proportion Correct", "
    Item Location","Item SE","Outfit MSE","Std. Outfit
    ", "Infit MSE","Std. Infit")

# Sort Table 2 by Item difficulty:
Table2 <- Table2[order(-Table2$`Item Location`),]

# Round the numeric values (all columns except the
    first one) to 2 digits:
Table2[, -1] <- round(Table2[,-1], digits = 2)
print.data.frame(Table2, row.names = FALSE)
```

##	Task ID	Proportion Correct	Item Location	Item SE
##	task_09	0.30	2.92	0.13
##	task_10	0.52	1.84	0.12
##	task_04	0.78	0.44	0.13
##	task_05	0.80	0.31	0.13
##	task_02	0.81	0.26	0.13
##	task_07	0.84	-0.02	0.14
##	task_03	0.88	-0.42	0.16
##	task_01	0.94	-1.24	0.20
##	task_08	0.97	-1.91	0.26
##	task_06	0.97	-2.18	0.29

##	Outfit MSE	Std. Outfit	Infit MSE	Std. Infit
##	1.05	0.46	0.90	-1.73
##	0.87	-1.82	0.90	-2.41
##	1.28	2.44	1.16	2.27
##	0.93	-0.62	0.99	-0.11
##	1.13	1.07	1.09	1.19
##	0.65	-2.77	0.80	-2.47
##	0.58	-2.74	0.75	-2.69
##	0.66	-1.33	0.76	-1.62
##	0.34	-2.21	0.69	-1.57
##	0.18	-2.79	0.61	-1.78

Finally, Table 3 provides a summary of the person calibrations. When there is a relatively large person sample size, it may be more useful to present the results as they relate to individual persons or subsets of the person sample as they are relevant to the purpose of the analysis. In the following code, we create Table 3 and print the first 6 lines to the console.

```
# Calculate proportion correct for persons:
PersonPropCorrect <- apply(student.locations$X.ex, 1,
    mean)
```

```
# Combine person calibration results in a table:
Table3 <-
   cbind.data.frame(achievement.with.se$id,
                    PersonPropCorrect,
                    achievement.with.se$`Person
                       Parameter`,
                    achievement.with.se$person_se,
                    person.fit$p.outfitMSQ,
                    person.fit$p.outfitZ,
                    person.fit$p.infitMSQ,
                    person.fit$p.infitZ)

names(Table3) <- c("Person ID", "Proportion Correct",
   "Person Location","Person SE","Outfit MSE","Std.
   Outfit", "Infit MSE","Std. Infit")

# Round the numeric values (all columns except the
   first one) to 2 digits:
Table3[, -1] <- round(Table3[,-1], digits = 2)
print.data.frame(Table3[1:6, ], row.names = FALSE)
```

```
##    Person ID Proportion Correct Person Location
##            1                0.8            1.93
##           10                0.6            1.16
##          100                0.4            1.16
##          101                0.7            3.03
##          102                0.9            0.52
##          103                0.7            3.03
##    Person SE Outfit MSE Std. Outfit Infit MSE
##         0.94       0.24       -0.52      0.41
##         0.83       1.10        0.36      1.13
##         0.83       5.62        3.50      2.10
##         1.19       0.56       -0.37      0.71
##         0.77       0.17        0.12      0.49
##         1.19       0.73       -0.09      0.80
##    Std. Infit
##         -1.19
##          0.47
##          2.55
##         -0.57
##         -0.61
##         -0.34
```

2.3   Dichotomous Rasch Model Analysis with MMLE in TAM

Next, we will use the TAM package (Robitzsch et al., 2021) to run the dichotomous Rasch model analyses. By default, the TAM package applies MMLE to estimate the dichotomous Rasch model. Please keep this estimation approach in mind when comparing the results between TAM and R packages or software programs that use other estimation techniques, such as Winsteps (Winsteps, 2020b) or Facets (Facets, 2020a).

Except where there are significant differences between the eRm and TAM procedures, we provide fewer details about the analysis procedures and interpretations in this section compared to the first illustration.

Prepare for the Analyses

To get started with the TAM package, install and load it into your R environment using the following code:

```
# install.packages("TAM")
library("TAM")
```

To facilitate the analysis, we will also use the *WrightMap* package (Irribarra and Freund, 2014):

```
# install.packages("WrightMap")
library("WrightMap")
```

If you have not already imported the Transitive Reasoning data and prepared it for analysis as described earlier in this chapter by isolating the response matrix, please do so before continuing with the TAM analyses.

Run the Dichotomous Rasch Model

Now, we are ready to run the dichotomous Rasch model on the transitive reasoning response data. We will use the `tam.mml()` function to run the model and store the results in an object called `dichot.transreas`. We specified `constraint = items` to obtain item locations that are centered at zero logits for ease of interpretation.

```
dichot.transreas_MMLE <- tam.mml(transreas.responses,
    constraint = "items", verbose = FALSE)
```

For brevity, we do not include all of the output from the model function in this book. However, after you run the model, you should see some output in the R console that includes information about each iteration in the estimation process.

Overall Model Summary

The next step is to request a summary of the model estimation results to begin to understand the results from the analysis. We can do so by applying the summary() function to the model object.

```
summary(dichot.transreas_MMLE)
```

```
## -----------------------------------------------------
## TAM 3.7-16 (2021-06-24 14:31:37)
## R version 4.1.0 (2021-05-18) x86_64, darwin17.0 |
    nodename=XXXX | login=root
##
## Date of Analysis: XXXX
## Time difference of 0.07364583 secs
## Computation time: 0.07364583
##
## Multidimensional Item Response Model in TAM
##
## IRT Model: PCM2
## Call:
## tam.mml(resp = transreas.responses, constraint = "
    items", verbose = FALSE)
##
## -----------------------------------------------------
## Number of iterations = 72
## Numeric integration with 21 integration points
##
## Deviance = 3354.91
## Log likelihood = -1677.46
## Number of persons = 425
## Number of persons used = 425
## Number of items = 10
## Number of estimated parameters = 11
##      Item threshold parameters = 9
##      Item slope parameters = 0
##      Regression parameters = 1
##      Variance/covariance parameters = 1
##
## AIC = 3377  | penalty=22    | AIC=-2*LL + 2*p
## AIC3 = 3388 | penalty=33    | AIC3=-2*LL + 3*p
## BIC = 3421  | penalty=66.57 | BIC=-2*LL + log(n
    )*p
```

```
## aBIC = 3386   | penalty=31.56      | aBIC=-2*LL + log
   ((n-2)/24)*p  (adjusted BIC)
## CAIC = 3432   | penalty=77.57      | CAIC=-2*LL + [
   log(n)+1]*p  (consistent AIC)
## AICc = 3378   | penalty=22.64      | AICc=-2*LL + 2*p
   + 2*p*(p+1)/(n-p-1)  (bias corrected AIC)
## GHP = 0.39728      | GHP=( -LL + p ) / (#Persons *
   #Items)  (Gilula-Haberman log penalty)
##
## -------------------------------------------------------
## EAP Reliability
## [1] 0.503
## -------------------------------------------------------
## Covariances and Variances
##        [,1]
## [1,] 0.938
## -------------------------------------------------------
## Correlations and Standard Deviations (in the
   diagonal)
##        [,1]
## [1,] 0.968
## -------------------------------------------------------
## Regression Coefficients
##          [,1]
## [1,] 1.93558
## -------------------------------------------------------
## Item Parameters -A*Xsi
##        item   N     M xsi.item AXsi_.Cat1
## 1   task_01 425 0.941   -1.233     -1.233
## 2   task_02 425 0.809    0.235      0.235
## 3   task_03 425 0.885   -0.432     -0.432
## 4   task_04 425 0.784    0.418      0.418
## 5   task_05 425 0.802    0.286      0.286
## 6   task_06 425 0.974   -2.133     -2.133
## 7   task_07 425 0.845   -0.047     -0.047
## 8   task_08 425 0.967   -1.875     -1.875
## 9   task_09 425 0.301    2.938      2.938
## 10  task_10 425 0.520    1.842      1.842
##      B.Cat1.Dim1
## 1              1
## 2              1
## 3              1
## 4              1
## 5              1
## 6              1
```

```
## 7                     1
## 8                     1
## 9                     1
## 10                    1
##
## Item Parameters in IRT parameterization
##          item alpha    beta
## 1   task_01      1  -1.233
## 2   task_02      1   0.235
## 3   task_03      1  -0.432
## 4   task_04      1   0.418
## 5   task_05      1   0.286
## 6   task_06      1  -2.133
## 7   task_07      1  -0.047
## 8   task_08      1  -1.875
## 9   task_09      1   2.938
## 10  task_10      1   1.842
```

From these results, we suggest taking a quick look at the item parameters as reported in the table labeled *Item Parameters in IRT parameterization*.

Because we ran a Rasch model, the *alpha* (discrimination) parameters are fixed to a constant value of 1. The *beta* parameters are the item locations on the latent variable. We will explore the item locations in more detail later in the analysis.

Wright Map

As the next step in our analysis, we will use the WrightMap package to create a Wright Map from our model results. We will create the plot using the function `IRT.WrightMap()` on the model object (`dichot.transreas`). As before, we will set the option for displaying threshold labels as `FALSE`, because we are working with dichotomous data. We also used the `main.title =` option to customize the title of the plot.

```
IRT.WrightMap(dichot.transreas_MMLE, show.thr.lab=
    FALSE, main.title = "Transitive\ Reasoning\ Wright
    \ Map (MMLE)")
```

Transitive Reasoning Wright Map (MMLE)

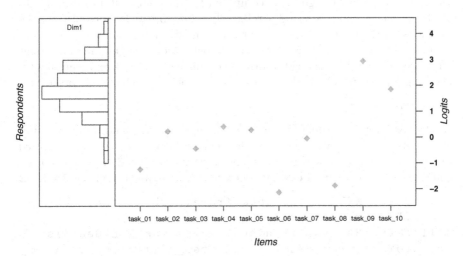

Items

In this *Wright Map* display, the results from the dichotomous Rasch model analysis of the transitive reasoning data are summarized graphically. The figure is organized as follows:

The left panel of the plot shows a histogram of respondent (person) locations on the logit scale that represents the latent variable. Units on the logit scale are shown on the far-right axis of the plot (labeled *Logits*). In the panel with the person locations, the label "Dim1" means that the distribution of person locations is specific to one dimension. In multidimensional Rasch models (Briggs and Wilson, 2003) and multidimensional IRT analyses (Bonifay, 2020; Reckase, 2009), the Wright Map can show multiple distributions of person locations that correspond to each dimension in the model.

The large central panel of the plot shows the item locations (item difficulty estimates) on the logit scale that represents the latent variable. Light grey diamond shapes show the logit-scale location of each item, as labeled on the x-axis.

Even though it is not appropriate to fully interpret item and person locations on the logit scale until there is evidence of acceptable model-data fit, we recommend examining the Wright Map during the preliminary stages of an item analysis to get a general sense of the model results and to identify any potential scoring or data entry errors.

A quick glance at the Wright Map suggests that, on average, the persons are located higher on the logit scale compared to the average item locations. In addition, there appears to be a relatively wide spread of person and item locations on the logit scale, such that the transitive reasoning test appears to be a useful tool for identifying differences in person locations and item locations on the latent variable. We will return to this display for more exploration after checking for acceptable model-data fit, among other psychometric properties.

Item Parameters

As the next step in our analysis, we will examine the item parameters in detail. We will use the $ operator to request the item location estimates that are stored in the item_irt table within the dichot.transreas_MMLE object. Then, we will save the item location estimates in a new object called difficulty_MMLE.

```
difficulty_MMLE <- as.data.frame(dichot.transreas_
    MMLE$item_irt$beta)
difficulty_MMLE$se <- dichot.transreas_MMLE$se.AXsi
names(difficulty_MMLE) <- c("item_difficulty", "item_
    se")
```

By running this code, we have created a data.frame object that includes two variables: item difficulty estimates and standard errors. The row labels show the item numbers from our response matrix. The item difficulty parameters and standard errors are interpreted in the same way as in the eRm analysis presented earlier in this chapter.

Next, let's calculate descriptive statistics to better understand the distribution of the item locations and standard errors. We will do so using the summary() function and the sd() function.

```
summary(difficulty_MMLE)
```

```
##   item_difficulty
##   Min.    :-2.13288
##   1st Qu.:-1.03275
##   Median :  0.09391
##   Mean    :  0.00000
##   3rd Qu.:  0.38541
##   Max.    :  2.93795
##        item_se.V1               item_se.V2
##   Min.    :0           Min.    :0.07786662
##   1st Qu.:0            1st Qu.:0.08245890
##   Median :0            Median :0.08679307
##   Mean    :0           Mean    :0.10578784
##   3rd Qu.:0            3rd Qu.:0.09815207
##   Max.    :0           Max.    :0.26524247
```

```
sd(difficulty_MMLE$item_difficulty)
```

```
## [1] 1.567331
```

```
sd(difficulty_MMLE$item_se)
```

```
## [1] 0.06678788
```

We can also visualize the item difficulty estimates in a simple histogram by using the hist() function. In our plot, we specified a custom title using main = and a custom x-axis label using xlab =.

```
hist(difficulty_MMLE$item_difficulty, main = "
    Histogram of Item Difficulty Estimates \nfor the
    Transitive Reasoning Data (MMLE)",
    xlab = "Item Difficulty Estimates in Logits")
```

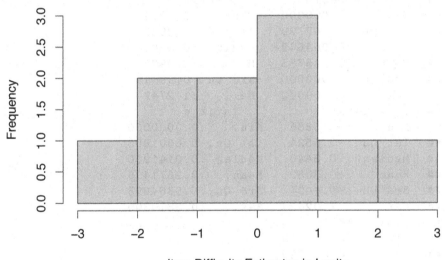

Item Fit

To calculate numeric item fit statistics, we will use the function tam.fit() from TAM on the model object (dichot.transreas_MMLE). We will store the item fit results in a new object called item.fit_MMLE, and then format this object as a data.frame for easy manipulation and exporting.

```
item.fit_MMLE <- tam.fit(dichot.transreas_MMLE)
item.fit_MMLE <- as.data.frame(item.fit_MMLE$itemfit)
```

Next, we will request a summary of the numeric fit statistics using the **summary** () function:

```
summary(item.fit_MMLE)
```

```
##     parameter                  Outfit
##   Length:9            Min.    :0.8371
##   Class  :character   1st Qu.:0.8971
##   Mode   :character   Median :0.9682
##                       Mean    :0.9873
##                       3rd Qu.:0.9866
##                       Max.    :1.2683
##      Outfit_t               Outfit_p
##   Min.    :-5.5429   Min.    :0.0000000
##   1st Qu.:-2.4857    1st Qu.:0.0003898
##   Median :-0.7044    Median :0.0129284
##   Mean    :-0.4300   Mean    :0.2256396
##   3rd Qu.:-0.3580    3rd Qu.:0.4811606
##   Max.    : 6.8686   Max.    :0.7203291
##    Outfit_pholm               Infit
##   Min.    :0.000000   Min.    :0.8329
##   1st Qu.:0.002729    1st Qu.:0.8982
##   Median :0.064642    Median :0.9751
##   Mean    :0.445785   Mean    :0.9923
##   3rd Qu.:1.000000    3rd Qu.:0.9941
##   Max.    :1.000000   Max.    :1.2747
##      Infit_t                 Infit_p
##   Min.    :-5.6886   Min.    :0.0000000
##   1st Qu.:-2.4524    1st Qu.:0.0001893
##   Median :-0.5440    Median :0.0141900
##   Mean    :-0.3082   Mean    :0.2671405
##   3rd Qu.:-0.1497    3rd Qu.:0.5864695
##   Max.    : 7.0243   Max.    :0.8810309
##    Infit_pholm
##   Min.    :0.000000
##   1st Qu.:0.001325
##   Median :0.070950
##   Mean    :0.453581
##   3rd Qu.:1.000000
##   Max.    :1.000000
```

The `item.fit_MMLE` object includes mean square error (MSE) and standardized (t) versions of the outfit and infit statistics for Rasch models. TAM also reports a p value for the standardized fit statistics (`Outfit_p` and `Infit_p`), along with adjusted significance values (`Infit_pholm` and `Outfit_pholm`).

With the item centering constraint that we used, the TAM package does not report fit statistics for all ten items. We can find approximate item fit statistics for all ten items by running the model without the centering constraint and saving the fit statistics.

```
dichot.transreas_MMLE2 <- tam(transreas.responses,
    verbose = FALSE)
item.fit_MMLE2 <- tam.fit(dichot.transreas_MMLE2)
item.fit_MMLE2 <- as.data.frame(item.fit_MMLE2$
    itemfit)
```

Person Parameters

When the default MMLE method is used to estimate the dichotomous Rasch model in TAM, person parameters are calculated after the item locations are estimated. As a result, the person estimation procedure for this model requires additional iterations.

In the following code, we calculate person locations that correspond to our model using the `tam.wle()` function with the dichotomous Rasch model object (`dichot.transreas_MMLE`). We stored the results in a new data frame called `achievement_MMLE`.

```
achievement_MMLE <- as.data.frame(tam.wle(dichot.
    transreas_MMLE))
```

The `achievement_MMLE` data frame includes the following variables:

- *pid*: Person identification number (based on person ordering in the item response matrix)
- *N.items*: Number of scored item responses for each person
- *PersonScores*: Sum of scored item responses for each person
- *PersonMax*: Maximum possible score for item responses
- *theta*: Person location estimate on the logit scale that represents the latent variable.
- *error*: Standard error for person location estimate, as described earlier in this chapter.
- *WLE.rel*: Reliability of person separation statistic, as described earlier in this chapter.

Next, we will calculate descriptive statistics to better understand the distribution of the person estimates. We will do so using the `summary()` function with the `achievement_MMLE` object.

```
summary(achievement_MMLE)
```

```
##       pid             N.items       PersonScores
## Min.   :  1    Min.   :10    Min.   : 1.000
## 1st Qu.:107    1st Qu.:10    1st Qu.: 7.000
## Median :213    Median :10    Median : 8.000
```

```
## Mean    :213    Mean    :10    Mean    : 7.828
## 3rd Qu.:319    3rd Qu.:10    3rd Qu.: 9.000
## Max.    :425    Max.    :10    Max.    :10.000
##    PersonMax        theta              error
## Min.    :10    Min.    :-2.592    Min.    :0.7508
## 1st Qu.:10    1st Qu.: 1.058    1st Qu.:0.8191
## Median :10    Median : 1.772    Median :0.9172
## Mean    :10    Mean    : 1.880    Mean    :1.0227
## 3rd Qu.:10    3rd Qu.: 2.727    3rd Qu.:1.1106
## Max.    :10    Max.    : 4.283    Max.    :1.7628
##    WLE.rel
## Min.    :0.3082
## 1st Qu.:0.3082
## Median :0.3082
## Mean    :0.3082
## 3rd Qu.:0.3082
## Max.    :0.3082
```

We can also visualize the person location estimates in a simple histogram by using the hist() function. In our plot, we specified a custom title using main = and a custom x-axis label using xlab =.

```
hist(achievement_MMLE$theta, main = "Histogram of
    Person Achievement Estimates \nfor the Transitive
    Reasoning Data \n (MMLE)",
    xlab = "Person Achievement Estimates Logits",
    cex.main = .9)
```

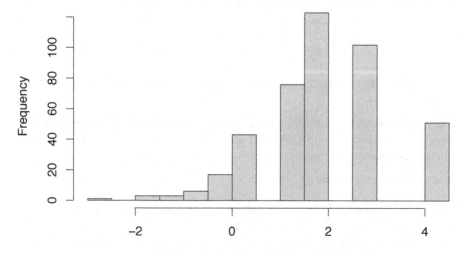

Histogram of Person Achievement Estimates for the Transitive Reasoning Data (MMLE)

Person Achievement Estimates in Logits

Person Fit

As a final step in our preliminary person analysis, we will conduct a brief exploration of person fit statistics. We explore person fit in more detail in Chapter 3.

To calculate numeric person fit statistics, we will use the function `tam.personfit()` from TAM on the model object (`dichot.transreas_MMLE`). We will store the person fit results in a new object called `person.fit_MMLE`, which is a data.frame object.

```
person.fit_MMLE <- tam.personfit(dichot.transreas_
    MMLE)
```

Next, we will request a summary of the numeric person fit statistics using the `summary()` function.

```
summary(person.fit_MMLE)
```

```
##    outfitPerson       outfitPerson_t
##   Min.   :0.04345    Min.    :-0.95709
##   1st Qu.:0.16335    1st Qu.:-0.63496
##   Median :0.46295    Median :-0.06839
##   Mean   :0.61001    Mean    : 0.05723
##   3rd Qu.:0.84557    3rd Qu.: 0.68920
##   Max.   :4.21347    Max.    : 2.84202
##    infitPerson        infitPerson_t
##   Min.   :0.1569     Min.    :-1.5008
##   1st Qu.:0.3613     1st Qu.:-0.9720
##   Median :0.6428     Median :-0.5789
##   Mean   :0.7734     Mean    :-0.2966
##   3rd Qu.:1.0647     3rd Qu.: 0.2983
##   Max.   :2.2003     Max.    : 2.5764
```

The `person.fit_MMLE` object includes mean square error (MSE) and standardized (t) versions of the person outfit and infit statistics for Rasch models.

2.4 Dichotomous Rasch Model Analysis with JMLE in TAM

We can also use the `tam.jml()` function to estimate the dichotomous Rasch model using JMLE, which is used in the Facets software program (Facets, 2020a) and the Winsteps software program (Winsteps, 2020b), which are popular standalone software programs for Rasch analyses.

With the exception of the model estimation code, most of the code for the JMLE approach is the same as the previous analysis with MMLE. Accordingly, we provide fewer explanations in our presentation of this code.

Estimate the Dichotomous Rasch Model

We already prepared the transitive reasoning data for analysis when we checked the data and isolated the item response matrix (`transreas.responses`) earlier in this chapter. We can begin our analysis using the `tam.jml()` function with this response matrix.

```
dichot.transreas_JMLE <- tam.jml(transreas.responses,
    constraint = "items", verbose = FALSE)
```

Overall Model Summary

```
# Request a summary of the model results
summary(dichot.transreas_JMLE)
```

Wright Map

When the JMLE estimation method is used, we need to specify the item location estimates and the person location estimates as vectors in the plotting function for the Wright Map. We will also use a different function to plot the Wright Map: `wrightMap()`, which takes arguments for the item and person locations.

In the following code, we saved the item locations in a vector called `difficulty_JMLE`, and the person locations in a vector called `theta_JMLE`. Then, we used these vectors as arguments in the `wrightMap()` function to create the Wright Map.

```
difficulty_JMLE <- dichot.transreas_JMLE$item$xsi.
    item
theta_JMLE <- dichot.transreas_JMLE$theta
wrightMap(thetas = theta_JMLE, thresholds =
    difficulty_JMLE, show.thr.lab=FALSE, main.title =
    "Transitive\ Reasoning\ Wright\ Map:\ JMLE")
```

Transitive Reasoning Wright Map: JMLE

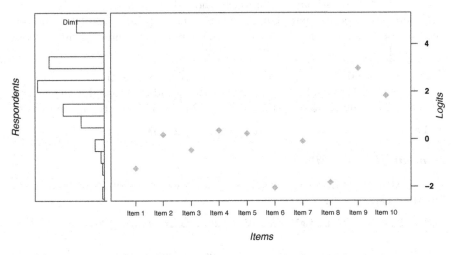

```
##                    [,1]
##    [1,]  -1.22764983
##    [2,]   0.19372234
##    [3,]  -0.46058464
##    [4,]   0.37626060
##    [5,]   0.24499876
##    [6,]  -2.04579773
##    [7,]  -0.08402885
##    [8,]  -1.81784893
##    [9,]   2.99257136
##   [10,]   1.82835693
```

Item Parameters

When we plotted the Wright Map, we created an object with the item difficulty parameters. We can find the standard errors for the item estimates by using the $ operator to extract errorP from the model results.

```
jmle.item.se <- dichot.transreas_JMLE$errorP
jmle.item.se
```

```
## [1] 0.10134872 0.08825844 0.09427051 0.08677937
## [5] 0.08783074 0.10718132 0.09071165 0.10578694
## [9] 0.08445690
```

Because of the item centering constraint that we included in our analysis, the standard error for the final item (item 10) is not reported.

Person Parameters

Unlike the MMLE estimation approach, the JMLE approach calculates item parameters and person parameters in the same step. As a result, we do not need to use a separate function to find the person parameter estimates from our model.

When we plotted the Wright Map, we already saved the theta estimates in a vector called `theta_JMLE`. We can find the standard errors for the person estimates by using the `$` operator to extract `errorWLE` from the model results.

```
jmle.person.se <- dichot.transreas_JMLE$errorWLE
```

Item and Person Fit

When JMLE is used, TAM calculates both person and item fit using a single function: `tam.jml.fit()`.

```
jmle.fit <- tam.jml.fit(dichot.transreas_JMLE)
```

Now that we have calculated fit statistics for items and persons, we will save the item fit statistics in a data frame called `jmle.item.fit`. We can examine the item fit statistics in more detail by printing the `jmle.item.fit` object to the console or preview window and calculating summary statistics.

```
jmle.item.fit <- as.data.frame(jmle.fit$fit.item)
print.data.frame(jmle.item.fit, row.names = FALSE)
```

```
##      item outfitItem outfitItem_t infitItem
##   task_01  0.4789486   -1.3892725 0.7618074
##   task_02  1.0961304    0.5264972 1.1485741
##   task_03  0.5213243   -1.9455120 0.7764883
##   task_04  1.2730392    1.4251000 1.2202071
##   task_05  0.8907676   -0.5080772 1.0439175
##   task_06  0.1379772   -2.0288954 0.5391578
##   task_07  0.6110629   -1.8512903 0.8437433
##   task_08  0.2378325   -1.7907958 0.6393029
##   task_09  0.8683358   -1.0803307 0.8585916
##   task_10  0.7954148   -2.2413359 0.8964146
##  infitItem_t
##    -1.6224741
##     1.8556014
##    -2.2436599
##     2.9097277
##     0.5973591
##    -2.3667045
##    -1.8383988
##    -1.9529903
##    -2.5264478
##    -2.1916767
```

```
summary(jmle.item.fit)
```

```
##        item              outfitItem
##   Length:10          Min.    :0.1380
##   Class :character   1st Qu.:0.4895
##   Mode  :character   Median :0.7032
##                      Mean    :0.6911
##                      3rd Qu.:0.8852
##                      Max.    :1.2730
##    outfitItem_t        infitItem
##   Min.    :-2.2413   Min.    :0.5392
##   1st Qu.:-1.9220    1st Qu.:0.7655
##   Median :-1.5900    Median :0.8512
##   Mean    :-1.0884   Mean    :0.8728
##   3rd Qu.:-0.6511    3rd Qu.:1.0070
##   Max.    : 1.4251   Max.    :1.2202
##    infitItem_t
##   Min.    :-2.5264
##   1st Qu.:-2.2307
##   Median :-1.8957
##   Mean    :-0.9380
##   3rd Qu.: 0.0424
##   Max.    : 2.9097
```

Next, we will save the person fit statistics in a data frame called `jmle.person`
`.fit`. We can examine the person fit statistics in more detail by viewing it in
the preview window and calculating summary statistics.

```
jmle.person.fit <- as.data.frame(jmle.fit$fit.person)
summary(jmle.person.fit)
```

```
##    outfitPerson        outfitPerson_t
##   Min.    :0.02708   Min.    :-0.9354
##   1st Qu.:0.17379    1st Qu.:-0.5101
##   Median :0.49116    Median : 0.2006
##   Mean    :0.69108   Mean    : 0.2507
##   3rd Qu.:0.89769    3rd Qu.: 0.7786
##   Max.    :4.95982   Max.    : 3.0980
##    infitPerson         infitPerson_t
##   Min.    :0.1067    Min.    :-1.51822
##   1st Qu.:0.4296     1st Qu.:-0.71666
##   Median :0.6515     Median :-0.31480
##   Mean    :0.8560    Mean    :-0.05799
##   3rd Qu.:1.1756     3rd Qu.: 0.56986
##   Max.    :2.3619    Max.    : 2.54349
```

2.5   Example Results Section

Table 2.1 presents a summary of the results from the analysis of the transitive
reasoning data (Sijtsma and Molenaar, 2002) using the dichotomous Rasch
model (Rasch, 1960). Specifically, the calibration of students ($N = 425$) and
tasks ($N = 10$) are summarized using average logit-scale calibrations, standard
errors, and model-data fit statistics. Examination of the results indicates that,
on average, the students were located higher on the logit scale ($M = 1.71$,
$SD = 1.09$), compared to tasks, whose locations were centered at zero logits
($M = 0.00$, $SD = 1.58$). This finding suggests that the items were relatively
easy for the sample of students who participated in this transitive reasoning
test. However, average values of the Standard Error (SE) were slightly larger
for students ($M = 0.95$) compared to Tasks ($M = 0.17$), indicating that
there may be some issues related to targeting for some of the students who
participated in the assessment. Average values of model-data fit statistics
indicate overall adequate fit to the model, with average outfit and infit mean
square statistics around 1.00, and average standardized outfit and infit statistics
near the expected value of 0.00 when data fit the model. This finding of overall
adequate fit to the model supports the interpretation of item and person
calibrations on the logit scale as indicators of their locations on the latent
variable measured by the test.

```
# Print Table 1:
tab1 <- knitr::kable(
  Table1, booktabs = TRUE,
  caption = 'Model Summary Table'
)
tab1 %>%
  kable_styling(latex_options = "scale_down", full_
      width = FALSE)
```

Table 2.2 includes detailed results for the 10 tasks included in the Transitive
Reasoning test. For each item, the proportion of correct responses is presented,
followed by the logit-scale location (δ), SE, and model-data fit statistics.
Examination of these results indicates that Task 9 was the most difficult
($ProportionCorrect = 0.30$; $\delta = 2.92$; $SE = 0.13$), followed by Task 10
($ProportionCorrect = 0.52$; $\delta = 1.84$; $SE = 0.12$). The easiest item was Task
6 ($ProportionCorrect = 0.97$; $\delta = -2.18$; $SE = 0.29$).

```
# Print Table 2:
tab2 <- knitr::kable(
  Table2, booktabs = TRUE,
  caption = 'Item Calibration'
)
```

TABLE 2.1
Model Summary Table

Statistic	Items	Persons
Logit Scale Location Mean	0.00	1.71
Logit Scale Location SD	1.58	1.09
Standard Error Mean	0.17	0.95
Standard Error SD	0.06	0.16
Outfit MSE Mean	0.77	0.77
Outfit MSE SD	0.35	0.80
Infit MSE Mean	0.86	0.91
Infit MSE SD	0.18	0.47
Std. Outfit Mean	−1.03	0.08
Std. Outfit SD	1.83	0.64
Std. Infit Mean	0.86	−0.12
Std. Infit SD	1.67	0.91
Reliability of Separation	1.00	0.21

TABLE 2.2
Item Calibration

Task ID	Proportion Correct	Item Location	Item SE	Outfit MSE	Std. Outfit	Infit MSE	Std. Infit
task_09	0.30	2.92	0.13	1.05	0.46	0.90	−1.73
task_10	0.52	1.84	0.12	0.87	−1.82	0.90	−2.41
task_04	0.78	0.44	0.13	1.28	2.44	1.16	2.27
task_05	0.80	0.31	0.13	0.93	−0.62	0.99	−0.11
task_02	0.81	0.26	0.13	1.13	1.07	1.09	1.19
task_07	0.84	−0.02	0.14	0.65	−2.77	0.80	−2.47
task_03	0.88	−0.42	0.16	0.58	−2.74	0.75	−2.69
task_01	0.94	−1.24	0.20	0.66	−1.33	0.76	−1.62
task_08	0.97	−1.91	0.26	0.34	−2.21	0.69	−1.57
task_06	0.97	−2.18	0.29	0.18	−2.79	0.61	−1.78

```
tab2 %>%
  kable_styling(latex_options = "scale_down", full_
     width = FALSE)
```

Table 2.3 includes detailed results for first 10 students who participated in the Transitive Reasoning Test. For each student, the proportion of correct responses is presented, followed by their logit-scale location estimate (θ), SE, and model-data fit statistics.

```
# Print the first 10 rows of Table 3:
tab3 <- knitr::kable(
  head(Table3,10), booktabs = TRUE,
  caption = 'Person Calibration'
)
tab3 %>%
  kable_styling(latex_options = "scale_down", full_
    width = FALSE)
```

Figure 1.1 illustrates the calibrations of the Participants and Items on the logit scale that represents the latent variable. The calibrations shown in this figure correspond to the results presented in Tables 2.2 and 2.3 for tasks and students, respectively. Starting at the bottom of the figure, the horizontal axis (labeled *Latent Dimension*) is the logit scale that represents the latent variable. In this analysis, lower numbers indicate less transitive reasoning ability, and higher numbers indicate more transitive reasoning ability. The central panel of the figure shows item difficulty locations on the logit scale for the 10 transitive reasoning tasks that were included in the analysis; the y-axis for this panel shows the item labels. By default in eRm, the items are ordered according to their original order in the response matrix. The upper panel of the figure shows a histogram of person (in this case, student) location estimates on the logit scale. Small vertical lines on the x-axis of this histogram show the points on the logit scale at which information (variance) is maximized for the sample of persons and items in the analysis.

On average, the persons (students) are located higher on the logit scale compared to the average item (task) locations. In addition, there appears

TABLE 2.3
Person Calibration

	Person ID	Proportion Correct	Person Location	Person SE	Outfit MSE	Std. Outfit	Infit MSE	Std. Infit
1	1	0.8	1.93	0.94	0.24	−0.52	0.41	−1.19
2	10	0.6	1.16	0.83	1.10	0.36	1.13	0.47
3	100	0.4	1.16	0.83	5.62	3.50	2.10	2.55
5	101	0.7	3.03	1.19	0.56	−0.37	0.71	−0.57
6	102	0.9	0.52	0.77	0.17	0.12	0.49	−0.61
8	103	0.7	3.03	1.19	0.73	−0.09	0.80	−0.34
9	104	0.9	1.93	0.94	0.17	0.12	0.49	−0.61
10	105	0.7	1.16	0.83	0.43	−0.64	0.59	−0.96
11	106	0.9	1.93	0.94	0.17	0.12	0.49	−0.61
12	107	0.9	1.16	0.83	1.47	0.92	1.64	0.97

to be a relatively wide spread of person and item locations on the logit scale, such that the transitive reasoning test appears to be a useful tool for identifying differences in person locations and item locations on the latent variable.

```
plotPImap(dichot.transreas, main = "Transitive
    Reasoning Assessment Wright Map")
```

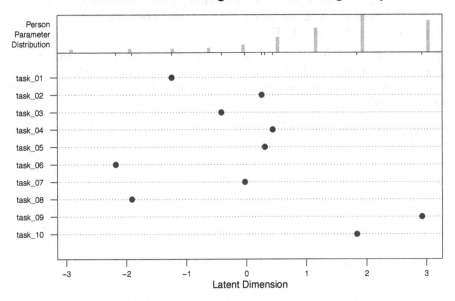

Transitive Reasoning Assessment Wright Map

2.6   Exercise

Estimate item and person locations with the dichotomous Rasch model for the Chapter 2 Exercise data using any of the estimation procedures included in this chapter. The Exercise 2 data include responses from 500 students to a mathematics achievement test that includes 35 items scored as incorrect ($x = 0$) or correct ($x = 1$). Then, try writing a results section similar to the example in this chapter to report your findings.

2.7   Supplementary Learning Materials

Rasch Estimation Demonstration Spreadsheet: https://www.rasch.org/moulton
.xls

Li, Y. Using the open-source statistical language R to analyze the dichotomous Rasch model. Behavior Research Methods 38, 532–541 (2006).
https://doi.org/10.3758/BF03192809

Rasch, G. (1960/1980). Probabilistic models for some intelligence and attainment tests.(Copenhagen, Danish Institute for Educational Research), expanded
edition (1980) with foreword and afterword by B.D. Wright. Chicago: The
University of Chicago Press.

Wright, B. D., & Masters, G. N. (1982). Rating Scale Analysis: Rasch Measurement. Chicago, IL: MESA Press.

3

Evaluating the Quality of Measures

This chapter provides an overview of several popular analyses that researchers can use to evaluate the quality of Rasch model results. Within the framework of Rasch measurement theory (Rasch, 1960), evaluating the quality of measures primarily involves exploring the degree to which the observed data (e.g., item responses) reflect fundamental measurement properties, as expressed in Rasch model requirements.

Procedures for evaluating measurement quality from a Rasch measurement perspective include techniques for evaluating *model-data fit*. Accordingly, this chapter begins with a brief overview of model-data fit analysis from the perspective of Rasch measurement theory (Rasch, 1960). Then, we demonstrate tools that analysts can use to evaluate model-data fit for Rasch models using R. As we did in Chapter 2, we use data from a transitive reasoning assessment (Sijtsma and Molenaar, 2002) to illustrate the application and interpretation of numeric and graphical model-data fit indicators. The chapter concludes with a challenge exercise and resources for further study.

We encourage readers to note that model-data fit analysis in modern measurement theory in general, as well as in the context of Rasch models in particular, is nuanced. The interpretation and use of model-data fit indices varies along with the individual purpose and consequences of each assessment. As a result, it is critical that analysts consider the unique context in which they are evaluating model-data fit when they interpret the results from model-data fit analyses, rather than relying exclusively on previously published critical values to interpret the results. In this chapter, we demonstrate techniques that analysts can use to evaluate model-data fit within the Rasch measurement theory framework. Our presentation is not exhaustive and there are many other methods that can be used to supplement those that we demonstrate here. We encourage interested readers to consult the resources listed throughout the chapter, as well as the resource list at the end of this chapter, to learn more about evaluating model-data fit within the context of Rasch measurement theory.

DOI: 10.1201/9781003174660-3

3.1 Evaluating Measurement Quality from the Perspective of Rasch Measurement Theory

When researchers evaluate model-data fit for Rasch models, they do so for the purpose of evaluating the psychometric quality of assessment procedures. In addition, such evaluations often include analyses related to issues of reliability, validity, and fairness. At the end of this chapter, we demonstrate techniques for calculating reliability indices using a Rasch approach. For additional discussions about these topics and demonstrations of how Rasch measurement indicators align with these foundational areas for evaluating psychometric procedures, we refer readers to the following resources:

- Engelhard, George. and Wind, Stefanie A. *Invariant measurement with raters and rating scales: Rasch models for rater-mediated assessments.* Taylor & Francis, 2018.

- Wolfe, Edward W. and Everrett V. Smith Jr. "Instrument development tools and activities for measure validation using Rasch models: Part I - instrument development tools." *Journal of Applied Measurement*, 8(1), 2007a, 97–123.

- Wolfe, Edward W. and Everrett V. Smith Jr. "Instrument development tools and activities for measure validation using Rasch models: Part II–validation activities." *Journal of Applied Measurement*, 8(2), 2007b, 204–234.

In the context of Rasch measurement theory (Rasch, 1960), model-data fit analyses focus on evaluating the item response patterns associated with individual items, persons, and other relevant elements of an assessment context (e.g., raters) for evidence of adherence to model requirements. This approach to model-data fit analysis helps researchers identify individual elements whose responses deviate from what would be expected given adherence to the Rasch model. Wright and Masters (1982) summarized this perspective as follows:

"The model is constructed to *govern our use of data* according to the characteristics we require of a measure and to show us, *through the exposure of discrepancies between intention and experience*, where our efforts to measure are threatened." (p. 90, emphasis added).

Deviations from model expectations can alert researchers to individual elements of an assessment context that warrant additional investigation (e.g., using qualitative analyses), revision (e.g., revisions to item text or revision of scoring materials in performance assessments), that can inform their theory about the construct, and that can help them identify areas for future research. This individual-element-focused approach to evaluating model-data fit in which model requirements are used as a framework against which to *evaluate* response patterns for the purpose of improving measurement procedures stands in contrast to model-data fit approaches whose goal is typically to *match* response patterns to a model whose parameters reflect them. This approach guides the analyses that we demonstrate in this book.

Residual-Based Fit Analyses

We focus on residual-based model-data fit indicators that are theoretically aligned with the Rasch measurement framework. Specifically, we demonstrate how analysts can calculate numeric and graphical summaries of *residuals*, or discrepancies between observed item responses and responses that would be expected given model estimates. In the framework of Rasch measurement theory, researchers typically examine residuals as they relate to individual items, persons, or other elements of the assessment context. These residual analyses provide insight into the degree to which the responses associated with elements of interest reflect model expectations.

Residuals (Y_{ni}) are calculated as follows:

$$Y_{ni} = X_{ni} - E_{ni} \tag{3.1}$$

where X_{ni} is the observed response from person n on item i, and E_{ni} is the model-expected value of the response for person n on item i.

To evaluate the magnitude of residuals, it is often useful to transform them to standardized residuals (Z_{ni}) as follows:

$$Z_{ni} = Y_{ni}/\sqrt{Wni} \tag{3.2}$$

where W_{ni} is the variance of the observed response.

For example, residual analyses could help a researcher identify an item in an attitude survey for which there were many unexpected responses. Further investigation of these responses using residuals could reveal that the unexpected responses occurred when participants who had otherwise generally negative attitudes, as reflected in low locations on the construct, gave higher-than-expected (i.e., more positive) responses to the item than would be expected given their location estimates. This information could reveal a potential issue related to the item content, which might be interpreted differently among participants with low locations on the construct. Alternatively, this information might provide insight into the nature of the construct and highlight a new direction for future research. Likewise, analysis of residuals associated with individual participants could help researchers identify participants who gave frequent or substantially unexpected responses to subsets of items.

We do not focus on global fit statistics that are typically reported in IRT analyses based on non-Rasch models for the purpose of comparing models (e.g., log-likelihood statistics), although these statistics are available in many of the R packages that include Rasch model analyses. Global fit and other overall model-data fit evaluation statistics are useful when the purpose of an analysis is to compare the fit of Rasch models to that of other candidate models; such analyses are beyond the scope of this book.

3.2   Example Data: Transitive Reasoning Test

We are going to practice evaluating model-data fit using the Transitive Reasoning assessment data that we explored in Chapter 2; these data were originally presented by Sijtsma and Molenaar (2002). Please see Chapter 2 for a detailed description of the data.

3.3   Rasch Model Fit Analysis with CMLE in *eRm*

We will use the *eRm* package (Mair et al., 2021) as the first package with which we demonstrate model-data fit analyses for Rasch models. We selected *eRm* as the first package for the analyses in the current chapter because it includes functions for evaluating model-data fit that are relatively straightforward to apply and interpret. Please note that the *eRm* package uses the Conditional Maximum Likelihood Estimation (CMLE) method to estimate Rasch model parameters. As a result, estimates from the *eRm* package are not directly comparable to estimates obtained using other estimation methods.

Prepare for the Analyses

We will prepare for the analyses by installing the *eRm* package and loading it into our R environment using the following code.

```
#install.packages("eRm")
library("eRm")
```

We will also use the *psych* package (Revelle, 2021) to perform a principal components analysis (PCA) of Rasch model standardized residual correlations. We install and load *psych* into our R environment using the following code.

```
# install.packages("psych")
library("psych")
```

Now that we have installed and loaded the packages to our R session, we are ready to import the data. We will use the function read.csv() to import the comma-separated values (.csv) file that contains the transitive reasoning response data. We encourage readers to use their preferred method for importing data files into R or R Studio.

Please note that if you use read.csv() you will need to first specify the file path to the location at which the data file is stored on your computer or set your working directory to the folder in which you have saved the data.

First, we will import the data using read.csv() and store it in an object called transreas.

```
transreas <- read.csv("transreas.csv")
```

In Chapter 2, we provided code to calculate descriptive statistics for the transitive reasoning data. We encourage readers to briefly explore their data using descriptive statistics before examining model-data fit.

Rasch Model Analysis

Because model-data fit analyses within the Rasch framework are based on residuals, the first step in the model-data fit analysis is to analyze the data with a Rasch model. Residuals are calculated using these estimates.

The transitive reasoning data in our example are scored responses in two categories ($x = 0$ or $x = 1$), and there are two major facets in the assessment system: students and items. Accordingly, we will apply the dichotomous Rasch model (Rasch, 1960) to the data to estimate student and item locations on the latent variable. Please refer to Chapter 2 of this book for more details about the dichotomous Rasch model.

First, we need to isolate the item response matrix from the other variables in the data (student IDs and grade level) so that we can analyze the responses using the dichotomous Rasch model. To do this, we will create an object made up of only the item responses. We will remove the first two variables (Student and Grade) from the transreas dataframe object using the subset() function with the select= option. We will save the resulting response matrix in a new dataframe object called transreas.responses.

```
transreas.responses <- subset(transreas, select = -c(
    Student, Grade))
```

Next, we will use summary() to calculate descriptive statistics for the transreas.responses object to check our work and ensure that the responses are ready for analysis.

```
summary(transreas.responses)
```

```
##       task_01                 task_02
##   Min.    :0.0000     Min.    :0.0000
##   1st Qu.:1.0000     1st Qu.:1.0000
##   Median :1.0000     Median :1.0000
##   Mean    :0.9412     Mean    :0.8094
##   3rd Qu.:1.0000     3rd Qu.:1.0000
##   Max.    :1.0000     Max.    :1.0000
##       task_03                 task_04
##   Min.    :0.0000     Min.    :0.0000
##   1st Qu.:1.0000     1st Qu.:1.0000
##   Median :1.0000     Median :1.0000
##   Mean    :0.8847     Mean    :0.7835
##   3rd Qu.:1.0000     3rd Qu.:1.0000
##   Max.    :1.0000     Max.    :1.0000
##       task_05                 task_06
##   Min.    :0.0000     Min.    :0.0000
##   1st Qu.:1.0000     1st Qu.:1.0000
##   Median :1.0000     Median :1.0000
##   Mean    :0.8024     Mean    :0.9741
##   3rd Qu.:1.0000     3rd Qu.:1.0000
##   Max.    :1.0000     Max.    :1.0000
##       task_07                 task_08
##   Min.    :0.0000     Min.    :0.0000
##   1st Qu.:1.0000     1st Qu.:1.0000
##   Median :1.0000     Median :1.0000
##   Mean    :0.8447     Mean    :0.9671
##   3rd Qu.:1.0000     3rd Qu.:1.0000
##   Max.    :1.0000     Max.    :1.0000
##       task_09                 task_10
##   Min.    :0.0000     Min.    :0.00
##   1st Qu.:0.0000     1st Qu.:0.00
##   Median :0.0000     Median :1.00
##   Mean    :0.3012     Mean    :0.52
##   3rd Qu.:1.0000     3rd Qu.:1.00
##   Max.    :1.0000     Max.    :1.00
```

Now, we are ready to run the dichotomous Rasch model on the transitive reasoning response data We will use the RM() function from *eRm* to run the model using CMLE and store the results in an object called dichot.transreas.

```
dichot.transreas <- RM(transreas.responses)
```

We will request a summary of the model results using the summary() function.

```
summary(dichot.transreas)
```

```
##
## Results of RM estimation:
##
## Call:  RM(X = transreas.responses)
##
## Conditional log-likelihood: -921.3465
## Number of iterations: 18
## Number of parameters: 9
##
## Item (Category) Difficulty Parameters (eta): with
   0.95 CI:
##          Estimate Std. Error lower CI upper CI
## task_02    0.258     0.133    -0.003    0.518
## task_03   -0.416     0.157    -0.723   -0.109
## task_04    0.441     0.128     0.190    0.692
## task_05    0.309     0.131     0.052    0.567
## task_06   -2.175     0.292    -2.747   -1.604
## task_07   -0.025     0.141    -0.302    0.252
## task_08   -1.909     0.262    -2.423   -1.395
## task_09    2.923     0.130     2.668    3.179
## task_10    1.836     0.115     1.610    2.062
##
## Item Easiness Parameters (beta) with 0.95 CI:
##                 Estimate Std. Error lower CI
## beta task_01     1.243      0.204     0.842
## beta task_02    -0.258      0.133    -0.518
## beta task_03     0.416      0.157     0.109
## beta task_04    -0.441      0.128    -0.692
## beta task_05    -0.309      0.131    -0.567
## beta task_06     2.175      0.292     1.604
## beta task_07     0.025      0.141    -0.252
## beta task_08     1.909      0.262     1.395
## beta task_09    -2.923      0.130    -3.179
## beta task_10    -1.836      0.115    -2.062
##                 upper CI
## beta task_01      1.643
## beta task_02      0.003
## beta task_03      0.723
## beta task_04     -0.190
## beta task_05     -0.052
## beta task_06      2.747
## beta task_07      0.302
## beta task_08      2.423
## beta task_09     -2.668
## beta task_10     -1.610
```

The summary of model results includes information about global model-data fit using a conditional log-likelihood statistic, the number of iterations used in the analysis, and the number of parameters. The summary also includes item *easiness* parameter estimates, which can be interpreted as the inverse of item difficulty parameters. The higher the value of this parameter, the easier the item is compared to the other items. This value can be multiplied by -1 to calculate item *difficulty* parameter estimates. Standard errors and 95% confidence intervals are also provided for the item location estimates.

At this point in the analysis, we will not explore the item and student calibrations in detail. Instead, we will focus on evaluating model-data fit. After we confirm adequate adherence to the Rasch model requirements, we can proceed to examine item and student estimates with confidence.

Evaluate Unidimensionality

As a first step in our analysis of the degree to which the transitive reasoning data adhere to the dichotomous Rasch model requirements, we will examine the Rasch model requirement of *unidimensionality*, or the degree to which there is evidence that one latent variable is sufficient to explain most of the variation in the responses.

The typical procedures for evaluating unidimensionality within the context of Rasch measurement theory are different from dimensionality assessment methods that researchers frequently use in psychometric analyses based on Classical Test Theory or non-Rasch IRT models (e.g., exploratory or confirmatory factor analyses). Specifically, the Rasch approach to exploring dimensionality begins with the requirement of unidimensionality as a prerequisite requirement for measurement. With this requirement in mind, the Rasch model is applied to the data to calculate estimates of item and person locations on a unidimensional scale. Then, model results are examined for evidence that one primary latent variable is sufficient to explain the variation in responses.

Our first step in evaluating adherence to the unidimensionality requirement is to calculate the proportion of variance in responses that can be attributed to item and person locations on the primary latent variable. To do this, we need to find the variance associated with the responses and the variance associated with residuals so that we can compare these values. We will start by calculating person parameters by applying the `person.parameter()` function to the model results object.

```
student.locations <- person.parameter(dichot.
    transreas)
```

Then, we can apply the `pmat()` function to the student locations object to calculate the model-predicted probabilities for a correct response for each student on each of the transitive reasoning tasks.

```
model.prob <- pmat(student.locations)
```

Because the *eRm* package uses CMLE to estimate student achievement parameters, the matrix of response probabilities excludes those students whose scored responses were extreme (in this case, all correct or all incorrect responses). As a result, we need to calculate residuals using a modified response matrix that does not include the students with extreme scores. This modified response matrix is stored in the `student.locations` object as `X.ex`. We will extract the response matrix from `student.locations` with the $ operator and store the result in a matrix called `responses.without.extremes`.

```
responses.without.extremes <- student.locations$X.ex
```

Now we can calculate residuals as the difference between the observed responses and model predictions.

```
resids <- responses.without.extremes - model.prob
```

Following Linacre (2003), we can calculate the proportion of variance in the responses associated with the Rasch model estimates using the following values:

- Variance of the observations: V_O
- Variance of the residuals: V_R
- Raw variance explained by Rasch measures: $(V_O - V_R)/V_O$

This procedure reveals the approximate proportion of variance in item responses that can be explained by the Rasch model estimates on the logit scale that represents the construct. This value is an approximation. Some researchers use a critical value of about 20% of variance explained by Rasch measures as evidence of adequate adherence to unidimensionality to support the interpretation of Rasch model results for many practical purposes (Reckase, 1979). We encourage readers to interpret this result as a continuous variable in the context of each unique assessment context. For more details and considerations on this topic, please see Linacre (2003). The following code calculates these values.

```
## Variance of the observations: VO
observations.vector <- as.vector(responses.without.
    extremes)
VO <- var(observations.vector)

## Variance of the residuals: VR
residuals.vector <- as.vector(resids)
VR <- var(residuals.vector)
```

```
## Raw variance explained by Rasch measures: (VO - VR
    )/VO
(VO - VR)/VO
```

```
## [1] 0.4115578
```

```
# Express the result as a percent:
((VO - VR)/VO) * 100
```

```
## [1] 41.15578
```

Our analysis indicates that approximately 41.16% of the variance in student responses to the transitive reasoning tasks can be explained by the Rasch model estimates of student and item locations on the logit scale that represents the latent variable.

Principal Components Analysis of Standardized Residual Correlations

It is also somewhat common in Rasch analysis to examine correlations among standardized residuals for evidence of potentially meaningful additional dimensions (i.e., factors, constructs, latent variables) beyond the primary latent variable that may be contributing to variance in responses (Linacre, 1998). When researchers use Rasch measurement theory to guide their analyses, they may use these exploratory techniques to *evaluate* the degree to which the construct of interest can be represented as a unidimensional latent variable (Borsboom and Mellenbergh, 2004; Maul, 2020; Michell, 1999). Deviations from unidimensionality provide information that is valuable for a variety of purposes. For example, deviations from unidimensionality might indicate that the guiding theory about the construct needs to be reconsidered or revised, that revisions are needed to ensure that the measurement procedure reflects the intended construct as a quantitative unidimensional variable, or both. In other cases, researchers might use dimensionality assessment tools as a method for evaluating a hypothesis that there are multiple distinct constructs related to a particular domain, as reflected in responses to an assessment procedure.

Some researchers evaluate adherence to the Rasch model requirement of unidimensionality using a PCA of the correlations among standardized residuals from Rasch model analyses. The basic idea behind this analysis is that if the responses adequately meet the Rasch model requirement of unidimensionality, then there should be no meaningful patterns in the residuals (i.e., the error or unexplained variance). The PCA of residual correlations is a post-hoc analysis conducted after the Rasch model has already been used to estimate person and item locations on the primary latent variable. As a result, the interpretation of the PCA of residual correlations within the framework of Rasch measurement theory is different from a typical interpretation of PCA in statistical analyses. Specifically, PCA of standardized residuals from the Rasch model is conducted for the purpose of *evaluating model requirements*; it is not used as a standalone statistical procedure.

Essentially, PCA of standardized residual correlations is a method for evaluating the degree to which additional latent variables beyond the primary latent variable may have contributed to item responses. For this reason, researchers who use PCA of standardized residuals in the context of Rasch analyses typically describe the eigenvalues from the PCA as *contrasts*, because they reflect *contrasting patterns* of responses to the primary latent variable. Relatively low values of contrasts provide evidence that the unidimensionality requirement is sufficiently satisfied.

Several researchers have conducted simulation and real data studies related to techniques for interpreting the values of contrasts from PCA of standardized residuals, including methods that involve identifying empirical critical values using simulation. We refer readers to the following resources to learn more about this method and best practices for interpreting the results:

- Chou, Yeh-Tai, and Wen-Chung Wang. "Checking Dimensionality in Item Response Models With Principal Component Analysis on Standardized Residuals." *Educational and Psychological Measurement*, 70(5), 2010, 717–731. https://doi.org/10.1177/0013164410379322

- Raîche, Gilles "Critical eigenvalue sizes in standardized residual principal components analysis." *Rasch Measurement Transactions* 19, 2005, 1012.

- Smith Jr., Everett V. "Detecting and evaluating the impact of multidimensionality using item fit statistics and principal component analysis of residuals." *Journal of Applied Measurement*, 3(2), 2002, 205–231.

- Smith, Richard M. "A comparison of methods for determining dimensionality in Rasch measurement." *Structural Equation Modeling*, 3(1), 1996, 25–40.

In this chapter, we provide some basic guidance for conducting a PCA of standardized residual correlations and interpreting the results. In the manual for the Winsteps software program, which includes this residual PCA procedure, Linacre's (2016) provides the following general guidance for interpreting the contrasts:

"We are trying to falsify the hypothesis that the residuals are random noise by finding the component that explains the largest possible amount of variance in the residuals. This is the 'first contrast' (or first PCA component in the correlation matrix of the residuals). If the eigenvalue of the first contrast is small (usually less than 2.0) then the first contrast is at the noise level and the hypothesis of random noise is not falsified in a general way. If not, the loadings on the first contrast indicate that there are contrasting patterns in the residuals."

We recommend interpreting the value of the contrasts using Linacre's recommendations as a starting point. When possible, we encourage analysts to use simulation methods to identify critical values for interpreting the results. Regardless of the method for identifying and interpreting contrasts relative to critical values, it is essential to note that real data will never perfectly adhere to the unidimensionality requirement or any Rasch model requirement. Instead of treating the results as categorical (e.g., "unidimensional" or "multidimensional"), we urge readers to consider the degree to which the results from these analyses provide useful information about potential deviations from the guiding measurement theory that may be useful for improving the measurement procedure, the theory about the construct, or both.

First, we need to obtain a matrix with the standardized residuals from the model. The *eRm* package provides the standardized residuals as part of the result from the itemfit() function. We need to apply itemfit() to the student.locations object, and then extract the standardized residuals. In the following code, we store the standardized residuals in an object called std.resids.

```
item.fit <- itemfit(student.locations)
std.resids <- item.fit$st.res
```

Now we will use the pca() function from the *psych* package (Revelle, 2021) to conduct the PCA on the standardized residuals object. We will save the first five contrasts in a vector called contrasts and plot the values using a simple graphical display.

```
pca <- pca(std.resids, nfactors = ncol(transreas.
    responses), rotate = "none")

contrasts <- c(pca$values[1], pca$values[2], pca$
    values[3], pca$values[4], pca$values[5])

plot(contrasts, ylab = "Eigenvalues for Contrasts",
    xlab = "Contrast Number", main = "Contrasts from
    PCA Standardized Residual Correlations")
```

Contrasts from PCA of Standardized Residual Correlations

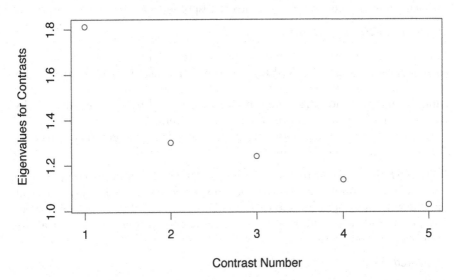

In this example, all of the contrasts have eigenvalues that are smaller than Linacre's (2016) suggested maximum value of 2.00. This result suggests that the correlations among the model residuals primarily reflect randomness (i.e., noise)—thus providing evidence that the responses adhere to the Rasch model requirement of unidimensionality.

Summaries of Residuals: Infit & Outfit Statistics

As a next step in evaluating the quality of our measurement procedure, we will calculate numeric fit statistics for the individual elements in the assessment procedure: items and persons. We will use popular Rasch fit statistics for practical purposes that are based on sums of squared residuals: Unweighted (outfit) and weighted (infit) mean square error (MSE) statistics. Unstandardized (χ^2) & standardized versions (Z or t) are available in most Rasch software programs. In this analysis, we will focus on the Unstandardized (χ^2) versions of these statistics.

Outfit Mean Square Error

Outfit MSE statistics are "unweighted fit" statistics. For items, outfit MSE is the sum of squared residuals for an item divided by the number of persons who responded to the item. For persons, outfit MSE is the sum of squared residuals for a person divided by the number of items encountered by the person.

Because it is an unweighted mean, the outfit statistic is sensitive to extreme departures from model expectations. For example, an extreme departure from model expectations would occur when an otherwise high-achieving student provided an incorrect response to a very easy item, or when an otherwise low-achieving student provided a correct response to a very difficult item.

Infit Mean Square Error

Infit stands for "information-weighted fit", where "information" means *variance*, such as larger variance for well-targeted observations, or smaller variance for extreme observations. For items, infit MSE statistics are calculated as the sum of squared standardized *item residuals*, weighted by variance, divided by the number of persons who responded to the item. For persons, infit MSE is the sum of squared standardized *person residuals*, weighted by variance, divided by the number of items the person encountered.

Infit MSE is sensitive to less-extreme unexpected responses compared to outfit MSE. Examples of less-extreme unexpected responses include a student providing an incorrect response to an item that is just below their achievement level, or a student providing a correct response to an item that is just above their achievement level.

Expected Values for MSE Fit Statistics

There is considerable disagreement among measurement scholars about how to classify an infit or outfit MSE statistic as evidence of "misfit" or "fit." Nonetheless, readers may find it useful to be aware of commonly agreed-upon principles for interpreting these statistics:

- The generally agreed-upon expected value is 1.00 when data fit the model

- Less than 1.00: Responses are more predictable than the model expects; responses resemble a Guttman-like (deterministic) pattern ("muted")

- Greater than 1.00: Responses are more haphazard ("noisy") than the model expects; there is too much variation to interpret that the estimate as a good representation of the response pattern

- Some variation is expected, but noisy responses are usually considered more cause for concern than muted responses

Item Fit Statistics

In practice, researchers usually begin individual Rasch fit analyses by examining fit statistics for items, rather than persons. There are two main reasons for this sequence. First, there is usually more data available with which to evaluate each item compared to each person, so the interpretation of fit statistics is less likely to be influenced by random fluctuations (i.e., noise). Second, it is usually easier to manage item misfit during a routine analysis compared to person misfit. For example, content validity concerns notwithstanding, in the case of severe item misfit, it may be possible to remove an item from an analysis and re-estimate parameters; removing persons from an analysis would likely warrant relatively more consideration. Alternatively, evidence of item misfit could signal data quality issues such as scoring or coding errors that could be resolved prior to additional analyses.

The *eRm* package includes the `itemfit()` function, which calculates Rasch model infit MSE and outfit MSE statistics for items. This function is applied to the object that includes person parameters. Earlier in this chapter, we applied `itemfit()` to the `student.locations` object and we stored the results in an object called `item.fit`. We will examine the `item.fit` object in more detail now by printing the results to our console:

```
item.fit
```

```
##
## Itemfit Statistics:
##              Chisq  df p-value Outfit MSQ Infit MSQ
## task_01 245.550 373   1.000      0.657     0.761
## task_02 421.904 373   0.041      1.128     1.087
## task_03 218.155 373   1.000      0.583     0.745
## task_04 478.894 373   0.000      1.280     1.158
## task_05 346.015 373   0.839      0.925     0.991
## task_06  65.613 373   1.000      0.175     0.611
## task_07 244.494 373   1.000      0.654     0.804
## task_08 125.478 373   1.000      0.336     0.687
## task_09 394.207 373   0.216      1.054     0.904
## task_10 326.534 373   0.960      0.873     0.898
##          Outfit t Infit t Discrim
## task_01    -1.327  -1.622   0.455
## task_02     1.071   1.187   0.054
## task_03    -2.742  -2.688   0.583
## task_04     2.442   2.268  -0.014
## task_05    -0.625  -0.109   0.184
## task_06    -2.789  -1.781   0.620
## task_07    -2.768  -2.475   0.493
## task_08    -2.205  -1.569   0.563
## task_09     0.462  -1.727   0.022
## task_10    -1.822  -2.408   0.226
```

The resulting table includes infit MSE and outfit MSE statistics for each item (labeled as "MSQ" statistics in the *eRm* package). The table also includes standardized versions of infit and outfit in the form of t statistics, along with an estimate of item discrimination (i.e., item slope). Item discrimination is not a parameter in the Rasch model, but it is calculated here as an additional indicator of fit to the Rasch model for individual items.

Person Fit Statistics

Next we will calculate numeric person fit statistics in the form of infit MSE and outfit MSE. In the *eRm* package, `personfit()` can be applied to the object with person parameter estimates to calculate these statistics. In our example, we will apply `personfit()` to the `student.locations` object and save the results in an object called `person.fit`. Because we have a relatively large sample, we will use the `summary()` function to explore the infit and outfit statistics rather than printing the results to the console like we did for the item fit statistics.

```
person.fit <- personfit(student.locations)

summary(person.fit$p.infitMSQ)

##     Min. 1st Qu.  Median    Mean 3rd Qu.    Max.
##   0.4060  0.4910  0.7487  0.9102  1.1740  2.2483

summary(person.fit$p.outfitMSQ)

##     Min. 1st Qu.  Median    Mean 3rd Qu.    Max.
##   0.1723  0.2399  0.5468  0.7665  0.9592  7.3060
```

These results indicate some variability in person fit, with infit MSE statistics ranging from 0.41 to 2.25, and outfit MSE statistics ranging from 0.17 to 7.31.

For discussions on interpreting person fit statistics in the context of Rasch measurement theory, we refer readers to the following resources:

- Smith, Richard. M. (1986). "Person Fit in the Rasch Model." *Educational and Psychological Measurement*, 46(2), 359–372. https://doi.org/10.1177/001316448604600210

- Walker A., Adrienne, Jeremy Kyle Jennings and George Engelhard. "Using person response functions to investigate areas of person misfit related to item characteristics." *Educational Assessment*, 23(1), 2018, 47–68. https://doi.org/10.1080/10627197.2017.1415143

- Wolfe W., Edward. "A bootstrap approach to evaluating person and item fit to the Rasch model." *Journal of Applied Measurement*, 14(1), 2013, 1–9.

3.4   Graphical Displays for Evaluating Model-Data Fit

Next, we will use some graphical displays to examine model-data fit in more detail. In this chapter, we focus on graphical displays for evaluating model-data fit specific to individual items. However, it is possible to create similar plots for persons as well as for other facets in an assessment context, such as raters.

Graphical displays are useful for exploring deviations from model expectations in more detail than is possible using the summary-level numeric fit statistics. In practice, many researchers use numeric fit statistics to identify individual items, persons, or other elements with notable misfit, and then use graphical displays to explore the nature of the misfit in more detail.

Plots of Standardized Residuals

First, we will create plots that show standardized residuals for the responses associated with individual items across persons. These plots demonstrate patterns in unexpected and expected responses that can be useful for understanding responses and interpreting results specific to individual items and persons.

Earlier in this chapter, we stored the standardized residuals in an object called std.resids. We will use this object to create plots for each of the individual items in the transitive reasoning assessment via a for-loop. For brevity, we have only included plots for the first three items here. The specific items to be plotted can be controlled by changing the items included in the items.to.plot object.

```
# Before constructing the plots, find the maximum and
    minimum values of the standardized residuals to
    set limits for the axes:
max.resid <- ceiling(max(std.resids))
min.resid <- ceiling(min(std.resids))


# The code below will produce plots of standardized
    residuals for selected items as listed in items.to
    .plot:
items.to.plot <- c(1:3)

for(item.number in items.to.plot){
  plot(std.resids[, item.number], ylim = c(min.resid,
      max.resid),
      main = paste("Standardized Residuals for Item
          ", item.number, sep = ""),
      ylab = "Standardized Residual", xlab = "Person
          Index")
  abline(h = 0, col = "blue")
  abline(h=2, lty = 2, col = "red")
  abline(h=-2, lty = 2, col = "red")

  legend("topright", c("Std. Residual", "Observed =
      Expected", "+/- 2 SD"), pch = c(1, NA, NA),
          lty = c(NA, 1, 2),
```

```
        col = c("black", "blue", "red"), cex = .8)
}
```

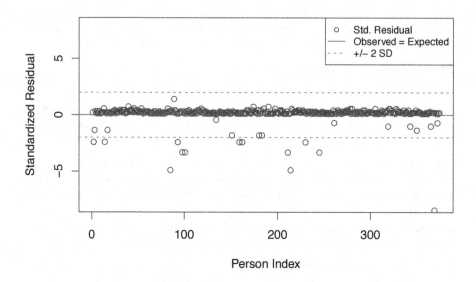

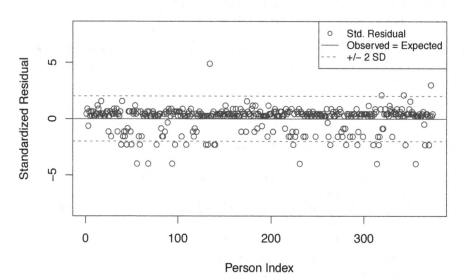

Standardized Residuals for Item 3

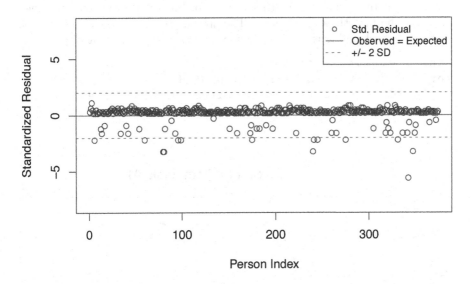

A separate plot is produced for each item. In each plot, the y-axis shows values of the standardized residuals, and the x-axis shows the persons (in this case, students), ordered by their relative position in the data set. Open-circle plotting symbols show the standardized residual associated with each person's response to the item of interest.

Horizontal lines are used to assist in the interpretation of the values of the standardized residuals. First, a solid line is plotted at a value of 0; standardized residuals equal to zero indicate that the observed response was equal to the model-expected response given person and item locations. Standardized residuals that are greater than zero indicate unexpected positive responses, and standardized residuals that are less than zero indicate unexpected negative responses. Dashed lines are plotted at values of +2 and −2 to indicate standardized residuals that are two standard deviations above or below model expectations, respectively. Researchers often interpret standardized residuals that exceed ±2 as evidence of statistically significant unexpected responses.

Empirical Item Response Functions

A second useful graphical display for evaluating model-data fit is a plot of model-expected item response functions (IRFs) overlaid with empirical (observed) IRFs (sometimes called item characteristic curves (ICCs)). These plots illustrate the relationship between the latent variable and the probability for a certain response (e.g., a correct response). Plotting the model-expected IRFs in the same display as the empirical IRFs provides insight into the frequency, magnitude, direction, and logit-scale location of expected and unexpected responses associated with individual items. To create these plots, we will apply the `plotICC()` function from the *eRm* package in a for-loop so that we can create plots for multiple items. For brevity, we have only included plots for the first three items here. The specific items to be plotted can be controlled by changing the items included in the `items.to.plot` object.

```
items.to.plot <- c(1:3)
for(item.number in items.to.plot){
    plotICC(dichot.transreas, item.subset = item.number
        , empICC = list("raw"), empCI = list())
}
```

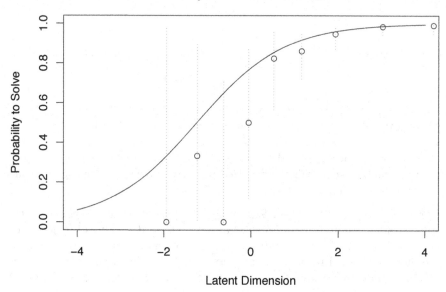

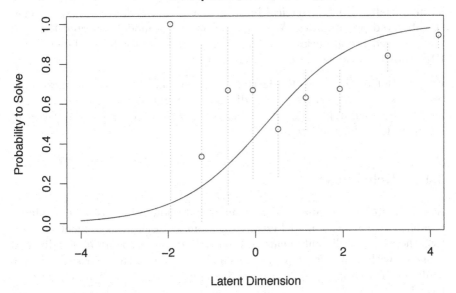

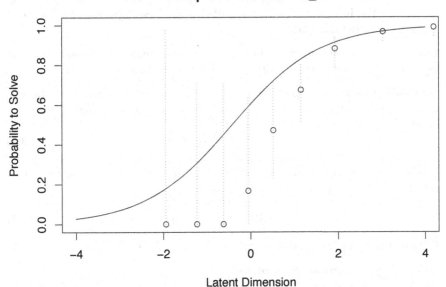

A separate plot is produced for each item. In each plot, the y-axis shows model-predicted probability for a correct or positive response (labeled "Probability to Solve" by default in the *eRm* package), and the x-axis shows the latent variable (labeled "Latent Dimension" by default in the *eRm* package). The solid line shows the model-expected IRF. Open-circle plotting symbols show the observed probability for a correct response at various locations on the latent variable. Dashed vertical lines show a 95% confidence interval around the observed probabilities. Alignment between the model-expected IRF and the observed probabilities (i.e., circles that are close to the model-expected IRF) provides evidence of good model-data fit.

Examination of these plots for the transitive reasoning items reveals unexpected responses that vary in frequency, direction, magnitude, and latent-variable location for each item. These results can be used to understand the nature of misfit in detail for each item.

3.4.1   Reliability

As a final step in the analyses, we demonstrate how it is possible to evaluate reliability using Rasch model estimates. From a Rasch measurement perspective, the focus in reliability analyses is on *ordering* and *separation* of individual elements within each facet (e.g., items and persons) on the logit scale. Specifically, the Rasch *reliability of separation statistic* can be calculated for each facet in the model (e.g., items and persons), and it is an estimate of *how well we can differentiate* individual items, persons, or other elements on the latent variable. We discussed the interpretation of this statistic in Chapter 2, and noted that the interpretation and use of this statistic depends on evidence of acceptable model-data fit.

Our transitive reasoning data include two facets: items and persons. Accordingly, we will calculate reliability of separation statistics for each facet.

Reliability of Item Separation The reliability of item separation statistic is calculated using a ratio of true (adjusted) variance to observed variance for items:

$$Rel_I = \left(SA_I^2\right) / \left(SD_I^2\right) \tag{3.3}$$

Where: $SA^2{}_I$ = Adjusted item variability; Calculated by subtracting error variance for items from total item variance:

$$SA_I^2 = SD_I^2 - SE_I^2 \tag{3.4}$$

$SD^2{}_I$ = Total item variance

Reliability of Person Separation

The reliability of person separation statistic is calculated using a ratio of true (adjusted) variance to observed variance for persons:

$$Rel_p = \left(SA_P^2\right) / \left(SD_P^2\right) \tag{3.5}$$

Where: $SA^2{}_P$ = Adjusted person variability; Calculated by subtracting error variance for persons from total person variance:

$$SA_P^2 = SD_P^2 - SE_P^2 \qquad (3.6)$$

$SD^2{}_P$ = Total person variance

The *eRm* package includes the function `SepRel()` to calculate the person separation reliability statistic. This function is applied to the person parameter object. We apply the `SepRel()` function to our `student.locations` object.

```
summary(SepRel(student.locations))
```

```
##
##          Separation Reliability:  0.2061
##
##              Observed Variance:  1.1784 (Squared
   Standard Deviation)
## Mean Square Measurement Error:  0.9355 (Model Error
   Variance)
```

3.5   Rasch Model Fit Analysis with MMLE in TAM

Next, we demonstrate model-data fit analyses using the *TAM* package (Robitzsch et al., 2021) with Marginal Maximum Likelihood Estimation (MMLE). The procedures for evaluating model-data fit with TAM are quite similar to those that we presented earlier in this chapter with *eRm*. As a result, we do not provide detailed annotations or discussions of the results from these analyses.

Prepare for the Analyses

```
# install.packages("TAM")
library("TAM")
```

Run the Rasch Model

```
dichot.transreas_MMLE <- tam.mml(transreas.responses,
   constraint = "items", verbose = FALSE)
```

Evaluate Unidimensionality

```
## Isolate the response matrix used in estimation:
resp <- dichot.transreas_MMLE$resp

## Find the expected response probabilities based on
   the model:
```

```
resids <- IRT.residuals(dichot.transreas_MMLE)

exp <- resids$X_exp

## Calculate raw (unstandardized) residuals:
resids.raw <- as.matrix(resp - exp)

## Calculate the variance in observations due to
   Rasch-model-estimated locations:

# Variance of the observations: VO
observations.vector <- as.vector(as.matrix(resp))
VO <- var(observations.vector)

# Variance of the residuals: VR
residuals.vector <- as.vector(resids.raw)
VR <- var(residuals.vector)

# Raw variance explained by Rasch measures: (VO - VR)
   / VO
(VO - VR)/VO

# Express the result as a percent:
((VO - VR)/VO) * 100
```

Principal Components Analysis of Standardized Residual Correlations

```
pca <- pca(resids$stand_residuals, nfactors = ncol(
   transreas.responses), rotate = "none")

contrasts <- c(pca$values[1], pca$values[2], pca$
   values[3], pca$values[4], pca$values[5])
plot(contrasts, ylab = "Eigenvalues for Contrasts",
   xlab = "Contrast Number", main = "Contrasts from
   PCA of Standardized Residual Correlations")
```

Contrasts from PCA of Standardized Residual Correlations

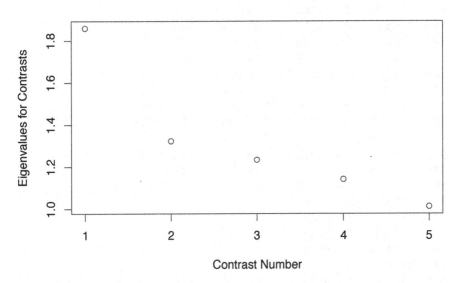

Numeric Fit Statistics

```
# Calculate numeric item fit statistics:
item.fit_MMLE <- tam.fit(dichot.transreas_MMLE)

## Item fit calculation based on 40 simulations
## |**********|
## |---------|

item.fit_MMLE <- as.data.frame(item.fit_MMLE$itemfit)
summary(item.fit_MMLE)

##    parameter              Outfit
##  Length:9           Min.   :0.8365
##  Class :character   1st Qu.:0.8973
##  Mode  :character   Median :0.9745
##                     Mean   :0.9902
##                     3rd Qu.:0.9898
##                     Max.   :1.2688
##    Outfit_t             Outfit_p
##  Min.   :-5.5464    Min.   :0.000000
##  1st Qu.:-2.4935    1st Qu.:0.000383
##  Median :-0.5676    Median :0.012650
##  Mean   :-0.3620    Mean   :0.259800
##  3rd Qu.:-0.2789    3rd Qu.:0.570305
##  Max.   : 6.9114    Max.   :0.780335
##    Outfit_pholm            Infit
```

```
##    Min.    :0.000000    Min.      :0.8321
##    1st Qu.:0.002681    1st Qu.:0.8971
##    Median :0.063249    Median :0.9779
##    Mean    :0.452261    Mean      :0.9915
##    3rd Qu.:1.000000    3rd Qu.:0.9922
##    Max.    :1.000000    Max.      :1.2695
##      Infit_t                Infit_p
##    Min.    :-5.7004    Min.      :0.0000000
##    1st Qu.:-2.4891    1st Qu.:0.0002199
##    Median :-0.4834    Median :0.0128067
##    Mean    :-0.3290    Mean      :0.2715080
##    3rd Qu.:-0.2049    3rd Qu.:0.6288306
##    Max.    : 6.9346    Max.      :0.8376471
##    Infit_pholm
##    Min.    :0.00000
##    1st Qu.:0.00154
##    Median :0.06403
##    Mean    :0.45254
##    3rd Qu.:1.00000
##    Max.    :1.00000
```

```
# Calculate numeric person fit statistics:
person.fit_MMLE <- tam.personfit(dichot.transreas_
    MMLE)
summary(person.fit_MMLE)
```

```
##    outfitPerson        outfitPerson_t
##    Min.    :0.04345    Min.      :-0.95709
##    1st Qu.:0.16335    1st Qu.:-0.63496
##    Median :0.46295    Median :-0.06839
##    Mean    :0.61001    Mean      : 0.05723
##    3rd Qu.:0.84557    3rd Qu.: 0.68920
##    Max.    :4.21347    Max.      : 2.84202
##    infitPerson        infitPerson_t
##    Min.    :0.1569    Min.      :-1.5008
##    1st Qu.:0.3613    1st Qu.:-0.9720
##    Median :0.6428    Median :-0.5789
##    Mean    :0.7734    Mean      :-0.2966
##    3rd Qu.:1.0647    3rd Qu.: 0.2983
##    Max.    :2.2003    Max.      : 2.5764
```

Graphical Fit Analysis

```
std.resids <- resids$stand_residuals

# Before constructing the plots, find the maximum and
    minimum values of the standardized residuals to
    set limits for the axes:
max.resid <- ceiling(max(std.resids))
min.resid <- ceiling(min(std.resids))

# The code below will produce standardized residual
    plots for each of selected items:
items.to.plot <- c(1:3)
for(item.number in items.to.plot){

  plot(std.resids[, item.number], ylim = c(min.resid,
     max.resid),
       main = paste("Standardized Residuals for Item
         ", item.number, sep = ""),
       ylab = "Standardized Residual", xlab = "Person
         Index")
  abline(h = 0, col = "blue")
  abline(h=2, lty = 2, col = "red")
  abline(h=-2, lty = 2, col = "red")

  legend("topright", c("Std. Residual", "Observed =
     Expected", "+/- 2 SD"), pch = c(1, NA, NA),
         lty = c(NA, 1, 2),
         col = c("black", "blue", "red"), cex = .8)
}
```

Standardized Residuals for Item 1

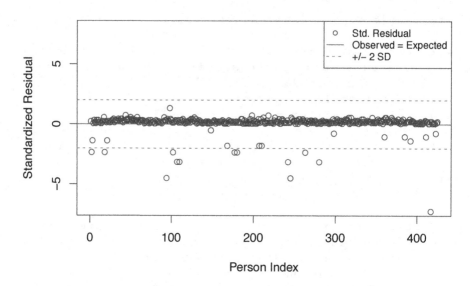

Standardized Residuals for Item 2

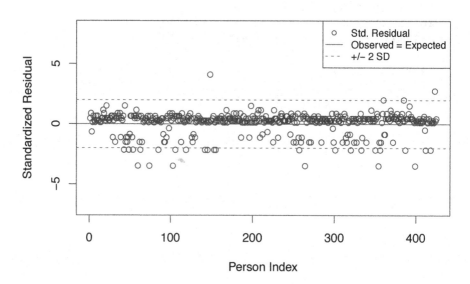

Standardized Residuals for Item 3

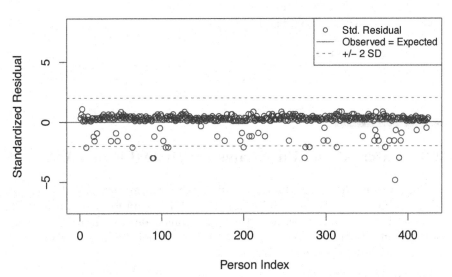

Reliability of Separation

```
## Person separation reliability:
dichot.transreas_MMLE$EAP.rel

## [1] 0.5030029

## Item separation reliability:

# Get Item scores
ItemScores <- colSums(transreas.responses)

# Get Item SD
ItemSD <- apply(transreas.responses,2,sd)

# Calculate the se of the Item
ItemSE <- ItemSD/sqrt(length(ItemSD))

# compute the Observed Variance (also known as Total
    Person Variability or Squared Standard Deviation)
SSD.ItemScores <- var(ItemScores)

# compute the Mean Square Measurement error (also
    known as Model Error variance)
Item.MSE <- sum((ItemSE)^2) / length(ItemSE)

# compute the Item Separation Reliability
```

```
item.separation.reliability <- (SSD.ItemScores-Item.
    MSE) / SSD.ItemScores
item.separation.reliability
```

```
## [1] 0.9999984
```

3.6 Rasch Model Fit Analysis with JMLE in TAM

Finally, we demonstrate model-data fit analyses using the *TAM* package
(Robitzsch et al., 2021) with Joint Maximum Likelihood Estimation (JMLE).
As in the previous section, we do not provide detailed annotations or discussions
of the results from these analyses because they are quite similar to those that
we have already presented in this chapter.

Run the Rasch Model

```
dichot.transreas_JMLE <- tam.jml(transreas.responses,
    verbose = FALSE)
```

Evaluate Unidimensionality

```
## Isolate the response matrix used in estimation:
resp <- dichot.transreas_JMLE$resp

## Find the expected response probabilities based on
    the model:
resids <- IRT.residuals(dichot.transreas_JMLE)

exp <- resids$X_exp

## Calculate raw (unstandardized) residuals:
resids.raw <- as.matrix(resp - exp)

## Calculate the variance in observations due to
    Rasch-model-estimated locations:

# Variance of the observations: VO
observations.vector <- as.vector(as.matrix(resp))
VO <- var(observations.vector)

# Variance of the residuals: VR
residuals.vector <- as.vector(resids.raw)
VR <- var(residuals.vector)
```

```
# Raw variance explained by Rasch measures: (VO - VR)
  / VO
(VO - VR)/VO
```

```
# Express the result as a percent:
((VO - VR)/VO) * 100
```

Principal Components Analysis of Standardized Residual Correlations

```
pca <- pca(resids$stand_residuals, nfactors = ncol(
    transreas.responses), rotate = "none")
```

```
contrasts <- c(pca$values[1], pca$values[2], pca$
    values[3], pca$values[4], pca$values[5])
plot(contrasts, ylab = "Eigenvalues for Contrasts",
    xlab = "Contrast Number", main = "Contrasts from
    PCA of Standardized Residual Correlations")
```

Contrasts from PCA of Standardized Residual Correlations

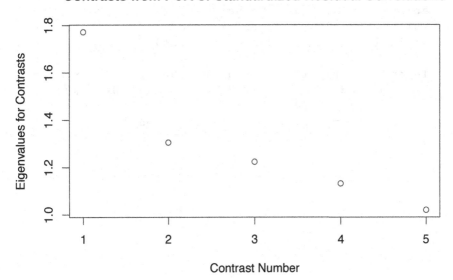

Item and Person Fit

```
fit.results <- tam.fit(dichot.transreas_JMLE)
```

```
item.fit <- fit.results$fit.item
person.fit <- fit.results$fit.person
```

Standardized Residual Plots

```
std.resids <- resids$stand_residuals

# Before constructing the plots, find the maximum and
    minimum values of the standardized residuals to
    set limits for the axes:
max.resid <- ceiling(max(std.resids))
min.resid <- ceiling(min(std.resids))

# The code below will produce standardized residual
    plots for each of the selected items:
items.to.plot <- c(1:3)
for(item.number in items.to.plot){

  plot(std.resids[, item.number], ylim = c(min.resid,
      max.resid),
      main = paste("Standardized Residuals for Item
          ", item.number, sep = ""),
      ylab = "Standardized Residual", xlab = "Person
          Index")
  abline(h = 0, col = "blue")
  abline(h=2, lty = 2, col = "red")
  abline(h=-2, lty = 2, col = "red")

  legend("topright", c("Std. Residual", "Observed =
      Expected", "+/- 2 SD"), pch = c(1, NA, NA),
          lty = c(NA, 1, 2),
          col = c("black", "blue", "red"), cex = .8)

}
```

Standardized Residuals for Item 1

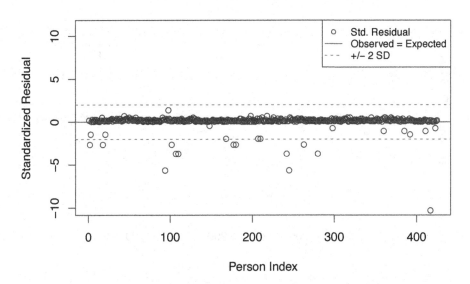

Standardized Residuals for Item 2

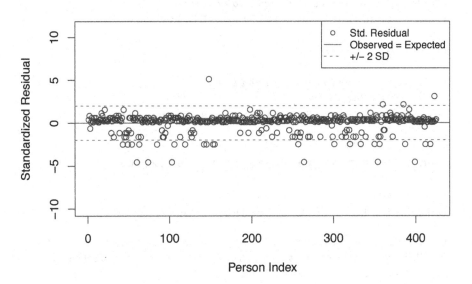

Standardized Residuals for Item 3

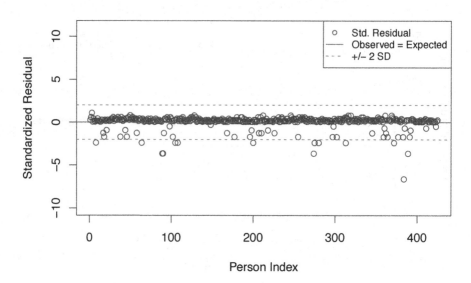

Person Index

Reliability of Separation

```
## Person separation reliability:
dichot.transreas_JMLE$WLEreliability

## [1] 0.3115775

## Item separation reliability:
# Get Item scores
ItemScores <- colSums(transreas.responses)

# Get Item SD
ItemSD <- apply(transreas.responses,2,sd)

# Calculate the se of the Item
ItemSE <- ItemSD/sqrt(length(ItemSD))

# compute the Observed Variance (also known as Total
    Person Variability or Squared Standard Deviation)
SSD.ItemScores <- var(ItemScores)

# compute the Mean Square Measurement error (also
    known as Model Error variance)
Item.MSE <- sum((ItemSE)^2) / length(ItemSE)

# compute the Item Separation Reliability
```

```
item.separation.reliability <- (SSD.ItemScores-Item.
    MSE) / SSD.ItemScores
item.separation.reliability
```

```
## [1] 0.9999984
```

3.7 Exercise

Evaluate model-data fit with the dichotomous Rasch model for the Chapter 3 Exercise data using the procedures included in this chapter. The Exercise 3 data are the same as the Exercise 2 data from Chapter 2; the data include responses from 500 students to a mathematics achievement test that includes 35 items scored as incorrect ($x = 0$) or correct ($x = 1$).

4

Rating Scale Model

This chapter provides a basic overview of the Rasch Rating Scale Model (RSM) (Andrich, 1978), along with guidance for analyzing data with the RSM using R. We use an example data set that includes participant responses to an attitude survey to illustrate the analysis using Conditional Maximum Likelihood Estimation (CMLE) via the *eRm* package (Mair et al., 2021). We also demonstrate RSM analyses using Marginal Maximum Likelihood Estimation (MMLE) and Joint Maximum Likelihood Estimation (JMLE) via the *TAM* package (Robitzsch et al., 2021). After the analyses are complete, we present an example description of the results. The chapter concludes with a challenge exercise.

Overview of the Rating Scale Model

David Andrich (1978) proposed the RSM (sometimes also called the Polytomous Rasch model) for use with ordinal item responses that are scored in more than two categories (e.g., data from attitude scales or performance assessments). Like the dichotomous Rasch model, the RSM provides estimates of *person locations* and *item locations* on a log-odds scale that represents the latent variable. The RSM also provides estimates of *rating scale category thresholds* that reflect the difficulty associated with each pair of adjacent categories in the rating scale. Specifically, the RSM specifies the probability for a rating in category k rather than in category $k-1$ as a function of the difference between person locations, item locations, and the location of the rating scale category threshold for category k on the logit scale.

The RSM can be stated in log-odds form as follows:

$$ln\left[\frac{P_{n_i(x=k)}}{P_{n_i(x=k-1)}}\right] = \theta_n - \delta_i - \tau_k \tag{4.1}$$

In the RSM, θ is the person's ability, δ is the item's difficulty, and τ is the rating scale category threshold. In the RSM, the threshold is the location on the logit scale at which there is an equal probability for a rating in category k and category $k-1$. For a rating scale made up of m categories, there are $m-1$ rating scale category thresholds.

The RSM produces one common set of rating scale category thresholds that apply to all of the items in the analysis. As a result, the RSM requires that the responses to all of the items include observations in the same categories. In addition, this common set of thresholds implies that the rating scale categories have a consistent interpretation across items. For example, for items with a four-category rating scale where 0 = *Strongly Disagree*, 1 = *Disagree*, 2 = *Agree*, and 3 = *Strongly Agree*, the RSM would produce a common set of three thresholds for all of the items. This common set of thresholds implies that the difference in the level of the latent variable required to provide a rating of *Strongly Agree* and *Agree* is the same for all of the items included in the analysis.

Rating Scale Model Requirements

Because it is a Rasch model, the RSM is based on the same requirements of unidimensionality, local independence, and invariance that we discussed in Chapter 2 for the dichotomous Rasch model. Evidence that rating scale responses approximate these requirements provides support for the meaningful interpretation and use of person, item, and threshold estimates on the logit scale as indicators of their respective ordering on the latent variable. In practice, many analysts evaluate some or all of these requirements using various indicators of model-data fit. In the current chapter, we provide code for calculating some popular residual-based fit indices for items and persons. Readers can use the same techniques that we considered in Chapter 3 to evaluate fit to the RSM.

4.1   Example Data: Liking for Science

The example data for this chapter are a group of 75 children's responses to the 25-item *Liking for Science* questionnaire, which was designed to measure their attitudes toward science activities. The data were published in Wright and Masters (1982). Each item stem included a science activity, and three response options: 0 = *Dislike*, 1 = *Not Sure/Don't Care*, and 2 = *Like*, such that responses in higher categories indicated more-favorable attitudes toward science activities.

4.2   RSM Analysis with CMLE in *eRm*

In the next section, we provide a step-by-step demonstration of a RSM analysis using the *eRm* package (Mair et al., 2021), which uses Conditional Maximum

Likelihood Estimation (CMLE). We encourage readers to use the example data set that is provided in the online supplement to conduct the analysis along with us.

Prepare for the Analyses

We selected *eRm* for the first illustration in the current chapter because it includes functions for applying the RSM that are relatively straightforward to use and interpret. Please note that the *eRm* package uses CMLE to estimate Rasch model parameters. As a result, estimates from the *eRm* package are not directly comparable to estimates obtained using other estimation methods.

First, install and load the *eRm* package into your R environment using the following code.

```
# install.packages("eRm")
library("eRm")
```

Next, we will use the function `read.csv()` to import the comma-separated values (.csv) file that contains the Liking for Science survey data. We encourage readers to use their preferred method for importing data files into R or R Studio.

Please note that if you use `read.csv()` you will first need to specify the file path to the location at which the data file is stored on your computer or set your working directory to the folder in which you have saved the data.

We will import the data using `read.csv()` and store it in an object called `science`.

```
science <- read.csv("liking_for_science.csv")
```

Next, we will explore the data using descriptive statistics using the `summary()` function.

```
summary(science)
##      student            i1              i2
##   Min.   : 1.0    Min.   :0.000   Min.   :0.000
##   1st Qu.:19.5    1st Qu.:1.000   1st Qu.:1.000
##   Median :38.0    Median :1.000   Median :2.000
##   Mean   :38.0    Mean   :1.453   Mean   :1.547
##   3rd Qu.:56.5    3rd Qu.:2.000   3rd Qu.:2.000
##   Max.   :75.0    Max.   :2.000   Max.   :2.000
##        i3              i4
##   Min.   :0.000   Min.   :0.0000
##   1st Qu.:1.000   1st Qu.:0.0000
##   Median :1.000   Median :1.0000
##   Mean   :1.173   Mean   :0.6933
```

```
## 3rd Qu.:2.000    3rd Qu.:1.0000
## Max.   :2.000    Max.    :2.0000
##          i5              i6              i7
## Min.   :0.0000    Min.    :0.000    Min.    :0.00
## 1st Qu.:0.0000    1st Qu.:1.000    1st Qu.:0.00
## Median :0.0000    Median :1.000    Median :1.00
## Mean   :0.4933    Mean    :1.213    Mean    :0.92
## 3rd Qu.:1.0000    3rd Qu.:2.000    3rd Qu.:1.50
## Max.   :2.0000    Max.    :2.000    Max.    :2.00
##          i8              i9             i10
## Min.   :0.00     Min.    :0.000    Min.    :0.000
## 1st Qu.:0.00     1st Qu.:0.000    1st Qu.:2.000
## Median :1.00     Median :1.000    Median :2.000
## Mean   :0.72     Mean    :1.067    Mean    :1.733
## 3rd Qu.:1.00     3rd Qu.:2.000    3rd Qu.:2.000
## Max.   :2.00     Max.    :2.000    Max.    :2.000
##         i11             i12             i13
## Min.   :0.000    Min.    :1.000    Min.    :0.000
## 1st Qu.:1.000    1st Qu.:2.000    1st Qu.:2.000
## Median :2.000    Median :2.000    Median :2.000
## Mean   :1.613    Mean    :1.827    Mean    :1.693
## 3rd Qu.:2.000    3rd Qu.:2.000    3rd Qu.:2.000
## Max.   :2.000    Max.    :2.000    Max.    :2.000
##         i14             i15             i16
## Min.   :0.000    Min.    :0.00     Min.    :0.000
## 1st Qu.:1.000    1st Qu.:1.00     1st Qu.:1.000
## Median :1.000    Median :2.00     Median :1.000
## Mean   :1.173    Mean    :1.48     Mean    :1.107
## 3rd Qu.:2.000    3rd Qu.:2.00     3rd Qu.:2.000
## Max.   :2.000    Max.    :2.00     Max.    :2.000
##         i17             i18             i19
## Min.   :0.000    Min.    :0.000    Min.    :0.00
## 1st Qu.:1.000    1st Qu.:2.000    1st Qu.:2.00
## Median :1.000    Median :2.000    Median :2.00
## Mean   :1.267    Mean    :1.933    Mean    :1.88
## 3rd Qu.:2.000    3rd Qu.:2.000    3rd Qu.:2.00
## Max.   :2.000    Max.    :2.000    Max.    :2.00
##         i20             i21             i22
## Min.   :0.0000    Min.    :0.000    Min.    :0.000
## 1st Qu.:0.0000    1st Qu.:1.000    1st Qu.:1.000
## Median :1.0000    Median :2.000    Median :1.000
## Mean   :0.6667    Mean    :1.587    Mean    :1.293
## 3rd Qu.:1.0000    3rd Qu.:2.000    3rd Qu.:2.000
## Max.   :2.0000    Max.    :2.000    Max.    :2.000
##         i23             i24             i25
```

```
##    Min.    :0.00    Min.     :0.000    Min.     :0.000
##    1st Qu.:0.00    1st Qu.:1.000    1st Qu.:1.000
##    Median :0.00    Median :2.000    Median :1.000
##    Mean    :0.56    Mean     :1.427    Mean     :1.133
##    3rd Qu.:1.00    3rd Qu.:2.000    3rd Qu.:2.000
##    Max.    :2.00    Max.     :2.000    Max.     :2.000
```

From the summary of `science`, we can see there are no missing data. We can also get a general sense of the scales, range, and distribution of each variable in the data set.

We can see that Student ID numbers range from 1 to 75, and that the maximum rating on the items was $x = 2$. Importantly, we can see that for item 12, the minimum rating was $x = 1$, and no ratings in the first category ($x = 0$) were observed in our sample. To use the RSM with these data, we will need to omit item 12 from our analysis. We will revisit these data in Chapter 5, where we include item 12 in a Partial Credit Model analysis. For now, we will create a new object called `science_drop_i12` that includes the Liking for Science data without item 12.

```
science_drop_i12 <- subset(science, select = -i12)
```

Run the Rating Scale Model

Next we need to isolate the item response matrix from the descriptive variables in the data (in this case, student IDs). To do this, we will create an object made up of only the item responses by removing the first variable (**student**) from the data.

```
science.responses <- subset(science_drop_i12, select
    = -student)
```

We will use `summary()` to calculate descriptive statistics for the `science.responses` object to check our work and ensure that the responses are ready for analysis:

```
summary(science.responses)
```

```
##            i1                  i2                  i3
##    Min.    :0.000    Min.     :0.000    Min.     :0.000
##    1st Qu.:1.000    1st Qu.:1.000    1st Qu.:1.000
##    Median :1.000    Median :2.000    Median :1.000
##    Mean    :1.453    Mean     :1.547    Mean     :1.173
##    3rd Qu.:2.000    3rd Qu.:2.000    3rd Qu.:2.000
##    Max.    :2.000    Max.     :2.000    Max.     :2.000
##            i4                  i5
##    Min.    :0.0000    Min.     :0.0000
##    1st Qu.:0.0000    1st Qu.:0.0000
```

```
##   Median  :1.0000     Median  :0.0000
##   Mean    :0.6933     Mean    :0.4933
##   3rd Qu.:1.0000      3rd Qu.:1.0000
##   Max.    :2.0000     Max.    :2.0000
##          i6                i7                i8
##   Min.    :0.000     Min.    :0.00     Min.    :0.00
##   1st Qu.:1.000      1st Qu.:0.00      1st Qu.:0.00
##   Median  :1.000     Median  :1.00     Median  :1.00
##   Mean    :1.213     Mean    :0.92     Mean    :0.72
##   3rd Qu.:2.000      3rd Qu.:1.50      3rd Qu.:1.00
##   Max.    :2.000     Max.    :2.00     Max.    :2.00
##          i9               i10               i11
##   Min.    :0.000     Min.    :0.000    Min.    :0.000
##   1st Qu.:0.000      1st Qu.:2.000     1st Qu.:1.000
##   Median  :1.000     Median  :2.000    Median  :2.000
##   Mean    :1.067     Mean    :1.733    Mean    :1.613
##   3rd Qu.:2.000      3rd Qu.:2.000     3rd Qu.:2.000
##   Max.    :2.000     Max.    :2.000    Max.    :2.000
##         i13               i14               i15
##   Min.    :0.000     Min.    :0.000    Min.    :0.00
##   1st Qu.:2.000      1st Qu.:1.000     1st Qu.:1.00
##   Median  :2.000     Median  :1.000    Median  :2.00
##   Mean    :1.693     Mean    :1.173    Mean    :1.48
##   3rd Qu.:2.000      3rd Qu.:2.000     3rd Qu.:2.00
##   Max.    :2.000     Max.    :2.000    Max.    :2.00
##         i16               i17               i18
##   Min.    :0.000     Min.    :0.000    Min.    :0.000
##   1st Qu.:1.000      1st Qu.:1.000     1st Qu.:2.000
##   Median  :1.000     Median  :1.000    Median  :2.000
##   Mean    :1.107     Mean    :1.267    Mean    :1.933
##   3rd Qu.:2.000      3rd Qu.:2.000     3rd Qu.:2.000
##   Max.    :2.000     Max.    :2.000    Max.    :2.000
##         i19               i20               i21
##   Min.    :0.00      Min.    :0.0000   Min.    :0.000
##   1st Qu.:2.00       1st Qu.:0.0000    1st Qu.:1.000
##   Median  :2.00      Median  :1.0000   Median  :2.000
##   Mean    :1.88      Mean    :0.6667   Mean    :1.587
##   3rd Qu.:2.00       3rd Qu.:1.0000    3rd Qu.:2.000
##   Max.    :2.00      Max.    :2.0000   Max.    :2.000
##         i22               i23               i24
##   Min.    :0.000     Min.    :0.00     Min.    :0.000
##   1st Qu.:1.000      1st Qu.:0.00      1st Qu.:1.000
##   Median  :1.000     Median  :0.00     Median  :2.000
##   Mean    :1.293     Mean    :0.56     Mean    :1.427
##   3rd Qu.:2.000      3rd Qu.:1.00      3rd Qu.:2.000
```

```
##    Max.    :2.000    Max.    :2.00    Max.    :2.000
##           i25
##    Min.    :0.000
##    1st Qu.:1.000
##    Median :1.000
##    Mean    :1.133
##    3rd Qu.:2.000
##    Max.    :2.000
```

Now, we are ready to run the RSM on the Liking for Science response data. We will use the RSM() function to run the model and store the results in an object called RSM.science.

```
RSM.science <- RSM(science.responses, se = TRUE)
```

Overall Model Summary

Next, we will request a summary of the model results using the summary() function.

```
summary(RSM.science)
```

```
##
## Results of RSM estimation:
##
## Call:   RSM(X = science.responses, se = TRUE)
##
## Conditional log-likelihood: -1162.281
## Number of iterations: 22
## Number of parameters: 24
##
## Item (Category) Difficulty Parameters (eta): with
##    0.95 CI:
##          Estimate Std. Error lower CI  upper CI
## i2       -0.748      0.210     -1.159    -0.337
## i3        0.319      0.184     -0.041     0.679
## i4        1.578      0.199      1.188     1.968
## i5        2.210      0.223      1.772     2.648
## i6        0.215      0.185     -0.147     0.577
## i7        0.966      0.186      0.601     1.330
## i8        1.502      0.197      1.116     1.888
## i9        0.591      0.183      0.232     0.950
## i10      -1.491      0.253     -1.986    -0.996
## i11      -0.983      0.221     -1.416    -0.551
## i13      -1.306      0.240     -1.776    -0.836
## i14       0.319      0.184     -0.041     0.679
## i15      -0.533      0.201     -0.927    -0.138
```

```
## i16          0.489          0.183      0.131      0.848
## i17          0.075          0.186     -0.291      0.440
## i18         -3.104          0.452     -3.990     -2.217
## i19         -2.455          0.349     -3.139     -1.771
## i20          1.656          0.201      1.261      2.050
## i21         -0.886          0.216     -1.309     -0.463
## i22          0.003          0.187     -0.364      0.371
## i23          1.984          0.213      1.566      2.402
## i24         -0.371          0.196     -0.756      0.013
## i25          0.421          0.183      0.062      0.780
## Cat 2        1.566          0.133      1.304      1.828
##
## Item Easiness Parameters (beta) with 0.95 CI:
##                   Estimate Std. Error lower CI upper CI
## beta i1.c1        0.451          0.199      0.062      0.840
## beta i1.c2       -0.664          0.409     -1.466      0.137
## beta i2.c1        0.748          0.210      0.337      1.159
## beta i2.c2       -0.071          0.426     -0.905      0.764
## beta i3.c1       -0.319          0.184     -0.679      0.041
## beta i3.c2       -2.203          0.396     -2.979     -1.428
## beta i4.c1       -1.578          0.199     -1.968     -1.188
## beta i4.c2       -4.722          0.447     -5.597     -3.847
## beta i5.c1       -2.210          0.223     -2.648     -1.772
## beta i5.c2       -5.986          0.499     -6.963     -5.009
## beta i6.c1       -0.215          0.185     -0.577      0.147
## beta i6.c2       -1.996          0.395     -2.771     -1.221
## beta i7.c1       -0.966          0.186     -1.330     -0.601
## beta i7.c2       -3.498          0.412     -4.306     -2.689
## beta i8.c1       -1.502          0.197     -1.888     -1.116
## beta i8.c2       -4.570          0.441     -5.436     -3.705
## beta i9.c1       -0.591          0.183     -0.950     -0.232
## beta i9.c2       -2.749          0.400     -3.533     -1.964
## beta i10.c1       1.491          0.253      0.996      1.986
## beta i10.c2       1.416          0.501      0.433      2.398
## beta i11.c1       0.983          0.221      0.551      1.416
## beta i11.c2       0.401          0.444     -0.470      1.271
## beta i13.c1       1.306          0.240      0.836      1.776
## beta i13.c2       1.046          0.478      0.110      1.982
## beta i14.c1      -0.319          0.184     -0.679      0.041
## beta i14.c2      -2.203          0.396     -2.979     -1.428
## beta i15.c1       0.533          0.201      0.138      0.927
## beta i15.c2      -0.501          0.413     -1.310      0.308
## beta i16.c1      -0.489          0.183     -0.848     -0.131
## beta i16.c2      -2.545          0.398     -3.325     -1.765
## beta i17.c1      -0.075          0.186     -0.440      0.291
```

```
## beta i17.c2   -1.715    0.396   -2.491   -0.940
## beta i18.c1    3.104    0.452    2.217    3.990
## beta i18.c2    4.641    0.896    2.885    6.397
## beta i19.c1    2.455    0.349    1.771    3.139
## beta i19.c2    3.344    0.690    1.992    4.695
## beta i20.c1   -1.656    0.201   -2.050   -1.261
## beta i20.c2   -4.877    0.452   -5.763   -3.991
## beta i21.c1    0.886    0.216    0.463    1.309
## beta i21.c2    0.206    0.436   -0.648    1.061
## beta i22.c1   -0.003    0.187   -0.371    0.364
## beta i22.c2   -1.573    0.396   -2.349   -0.796
## beta i23.c1   -1.984    0.213   -2.402   -1.566
## beta i23.c2   -5.535    0.478   -6.472   -4.598
## beta i24.c1    0.371    0.196   -0.013    0.756
## beta i24.c2   -0.824    0.406   -1.619   -0.029
## beta i25.c1   -0.421    0.183   -0.780   -0.062
## beta i25.c2   -2.409    0.397   -3.187   -1.631
```

The summary of the RSM output includes the Conditional Log-likelihood statistic, details about the number of iterations and model parameters, and a table with item parameters, their standard errors, and confidence intervals. It is important to note that the item parameters included in this preliminary output are *item easiness* parameters—*not* item difficulty parameters. We will examine item difficulty parameters in detail later in our analysis.

Wright Map

We will create a Wright Map (see Chapter 2) from our model results. This display will provide an overview of the distribution of person, item, and threshold parameters. We will create the plot using the *eRm* package function plotPImap() on the model object RSM.science.

```
plotPImap(RSM.science, main = "Liking for Science
    Rating Scale Model Wright Map")
```

Liking for Science Rating Scale Model Wright Map

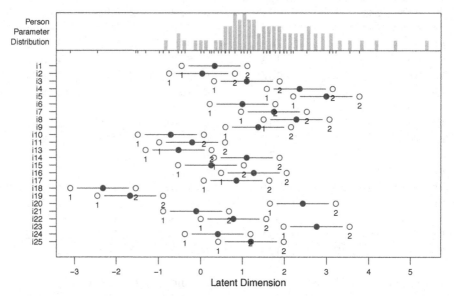

In this *Wright Map* display, the results from the RSM analysis of the Liking for Science data are summarized graphically. The figure is organized as follows.

Starting at the bottom of the figure, the horizontal axis (labeled *Latent Dimension*) is the logit scale that represents the latent variable. In the application of the Liking for Science data, lower numbers indicate less-favorable attitudes toward science activities, and higher numbers indicate more-favorable attitudes toward science activities.

The central panel of the figure shows item difficulty locations on the logit scale for the 24 Liking for Science items that were included in the analysis; the y-axis for this panel shows the item labels. By default in *eRm*, the items are ordered according to their original order in the response matrix. The items can be ordered by difficulty by adding `sorted = TRUE` as an argument in the `plotPImap()` call. For each item, a solid circle plotting symbol shows the overall location estimate. This solid circle symbol is connected to two open-circle symbols that show the locations of the rating scale category thresholds. Each threshold is labeled with a 1 to indicate the threshold between rating scale category $x = 0$ and $x = 1$, or a 2 to indicate the threshold between rating scale category $x = 1$ and $x = 2$.

The upper panel of the figure shows a histogram of person (in this case, children) location estimates on the logit scale. Small vertical lines on the x-axis of this histogram show the points on the logit scale at which information (variance) is maximized for the sample of persons and items in the analysis. These lines can be omitted by adding `irug = FALSE` as an argument in the `plotPImap()` call.

Item Parameters

Next, we will examine the item difficulty location and rating scale category threshold estimates. The *eRm* package provides several options with which analysts can find and examine item and rating scale category threshold parameters. For example, one way to obtain the overall item location parameters (δ) is to extract the *eta parameters* from the model object using the \$ operator. We extract these parameters, print them to the console, and calculate summary statistics for them with the following code.

```
item.locations <- RSM.science$etapar
item.locations
```

```
##               i2              i3              i4
## -0.747716237    0.318637705    1.578076249
##               i5              i6              i7
##   2.209910144    0.214960484    0.965680071
##               i8              i9             i10
##   1.502096325    0.591198710   -1.490900062
##              i11             i13             i14
## -0.983450547   -1.306083786    0.318637700
##              i15             i16             i17
## -0.532510932    0.489411243    0.074568234
##              i18             i19             i20
## -3.103553567   -2.454966823    1.655551546
##              i21             i22             i23
## -0.886223768    0.003185124    1.984230774
##              i24             i25           Cat 2
## -0.371230973    0.421330341    1.566148493
```

```
summary(item.locations)
```

```
##      Min.   1st Qu.    Median      Mean   3rd Qu.
## -3.10355  -0.78234   0.26680   0.08404   1.09978
##      Max.
##   2.20991
```

Because of the nature of the estimation process used in *eRm*, the `item.locations` object that we just created does not include the location estimate for the first item. One can calculate the location for item 1 by subtracting the sum of the item locations from zero. In the following code, we find the location for item 1, and then create a new object with all 24 item locations.

```
n.items <- ncol(science.responses)
i1 <- 0 - sum(item.locations[1:(n.items - 1)])
item.locations.all <- c(i1, item.locations[c(1:(n.
    items - 1))])
item.locations.all
```

```
##                           i2              i3
## -0.450837954  -0.747716237   0.318637705
##              i4              i5              i6
##  1.578076249   2.209910144   0.214960484
##              i7              i8              i9
##  0.965680071   1.502096325   0.591198710
##             i10             i11             i13
## -1.490900062  -0.983450547  -1.306083786
##             i14             i15             i16
##  0.318637700  -0.532510932   0.489411243
##             i17             i18             i19
##  0.074568234  -3.103553567  -2.454966823
##             i20             i21             i22
##  1.655551546  -0.886223768   0.003185124
##             i23             i24             i25
##  1.984230774  -0.371230973   0.421330341
```

Alternatively, one can apply the `thresholds()` function to the model object in order to find the item locations from the RSM. This procedure provides item location estimates (δ) as well as estimates of the item location combined with rating scale category thresholds ($\delta + \tau$). However, the values produced with this function are not centered at zero logits, so a little manipulation is required to obtain the centered values. In the following code chunk, we apply the `thresholds()` function to obtain the uncentered item location estimates and then calculate centered item locations.

```
# Apply thresholds() function to the model object in
    order to obtain item locations (not centered at
    zero logits):
items.and.taus <- thresholds(RSM.science)
items.and.taus.table <- as.data.frame(items.and.taus$
    threshtable)
uncentered.item.locations <- items.and.taus.table$X1.
    Location

# Set the mean of the item locations to zero logits:
centered.item.locations <- scale(uncentered.item.
    locations, scale = FALSE)
summary(centered.item.locations)
```

```
##           V1
##   Min.    :-3.1036
##   1st Qu.:-0.7823
##   Median : 0.1448
##   Mean   : 0.0000
```

```
##    3rd Qu.:  0.6848
##    Max.    : 2.2099
```

In addition, the eta parameters from the RSM object include *cumulative* rating scale category thresholds, printed after the final item location estimate. In our example, this value is labeled "Cat 2". Cumulative thresholds are not always used in Rasch model applications; Rasch-Andrich (i.e., adjacent categories) thresholds are often used instead (Andrich, 2015; Mellenbergh, 1995). We will calculate the values of the rating scale category thresholds using the results from the `thresholds()` function. This function produces estimates of the rating scale category thresholds combined with item difficulty parameter estimates. We can subtract these values from the item locations to find the values of each threshold.

In our example, we have a rating scale with three categories, so we have two rating scale category thresholds. In the following code chunk, we create an empty object in which to store the threshold estimates (`tau.estimates`), and then we use a for-loop to calculate the estimates and store them in our object. Because the thresholds are the same for all of the items, we will save the first value from vector of differences between the item+threshold locations and the overall locations.

```
# Specify the number of thresholds as the maximum
    observed score in the response matrix (be sure the
    responses begin at category 0):
n.thresholds <- max(science.responses)

# Calculate adjacent-category threshold values:
tau.estimates <- NULL

for(tau in 1:n.thresholds){
   tau.estimates[tau] <- (items.and.taus.table[, (1+
       tau)] - items.and.taus.table[,1])[1]
}
```

Next, we will calculate standard errors for each item and threshold location and store them in new objects. We will use `summary()` to examine descriptive statistics for these values.

```
#SE for items + thresholds:
delta.tau.se <- items.and.taus$se.thresh
summary(delta.tau.se)
```

```
##    Min. 1st Qu.  Median    Mean 3rd Qu.    Max.
##  0.1831  0.1989  0.2303  0.2379  0.2459  0.4535
```

```
# SE for overall item:
delta.se <- RSM.science$se.eta
summary(delta.se)
```

```
##     Min. 1st Qu.  Median    Mean 3rd Qu.    Max.
##   0.1335  0.1844  0.1980  0.2153  0.2172  0.4524
```

Item Response Functions

We will examine graphical displays of item difficulty using item response functions. With the RSM, the *eRm* package creates plots of the probability for a rating in each rating scale category, conditional on person locations on the latent variable. In the following code, we use `plotICC()` from *eRm* to create rating scale category probability plots. We included `ask = FALSE` in our function call in order to generate all of the plots at once. For brevity, we have only included plots for the first three items here. The specific items to be plotted can be controlled by changing the items included in the `items.to.plot` object.

```
items.to.plot <- c(1:3)
plotICC(RSM.science, ask = FALSE, item.subset = items
    .to.plot)
```

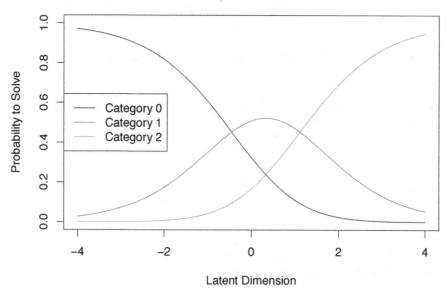

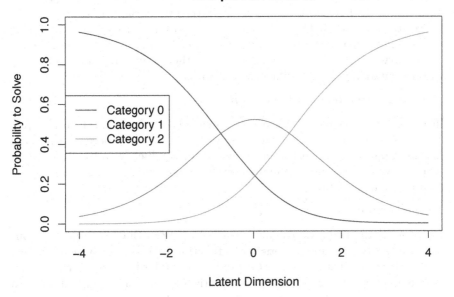

ICC plot for item i2

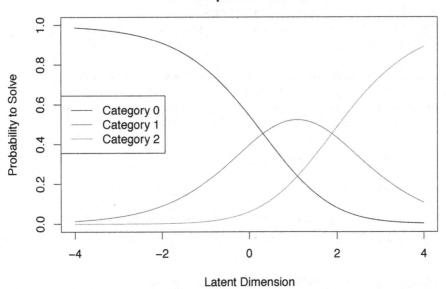

ICC plot for item i3

This code generates plots of rating scale category probabilities for individual items. In each item-specific plot, the x-axis is the logit scale that represents the latent variable; this scale represents favorability toward science activities in our example. The y-axis is the probability for a rating in each category, conditional on person locations on the latent variable. Separate lines in different colors show the conditional probability for each category of responses observed for the item of interest. Because we used the RSM, the overall shape of the curves and the relative distance between the curves is consistent across items—reflecting the common set of rating scale category thresholds. However, the location of the curves on the logit scale shifts across the individual items—reflecting each item's unique overall difficulty location (δ).

Person Parameters and Item Fit

In the *eRm* package, it is necessary to calculate person parameters *before* item fit statistics can be calculated. Accordingly, we will proceed with a brief examination of person parameters before we conduct item fit analyses. In practice, we recommend examining item fit before examining and interpreting item locations in detail.

Examine Person Parameters

In the following code, we calculate person locations that correspond to our model using the `person.parameter()` function with the RSM model object (`RSM.science`). This function also produces standard errors for the person locations. We stored the person location estimates and their standard errors in a new data frame called `person.locations`, and then requested a summary of the estimation results using `summary()`.

```
# Calculate person parameters:
person.locations.estimate <- person.parameter(RSM.
    science)

# Store person parameters and their standard errors
    in a dataframe object:
person.locations <- cbind.data.frame(person.locations
    .estimate$thetapar,

                                    person.locations
                                       .estimate$se.
                                       theta)
names(person.locations) <- c("theta", "SE")

# View summary statistics for person parameters:
summary(person.locations)

##         theta                    SE
##   Min.    :-0.8248    Min.     :0.3362
##   1st Qu.: 0.8173     1st Qu.:0.3377
##   Median : 1.3296     Median :0.3474
##   Mean    : 1.5669    Mean     :0.3819
##   3rd Qu.: 2.2379     3rd Qu.:0.3768
##   Max.    : 5.3886    Max.     :1.0285
```

The estimation procedure in *eRm* does not directly produce parameter estimates for persons with extreme scores. In our example, extreme scores would result from a child giving a response of $x = 0$ to all items or a child giving a response of $x = 2$ to all items. For these students, a standard error is not calculated. In our example, Child #2 had an extreme score because they gave a rating in category 2 to all items.

Examine Item Fit

Next, we will conduct a brief exploration of item fit statistics for the Liking for Science items. We considered item fit in detail in Chapter 3; readers can use the procedures in that chapter to examine item fit in detail for the RSM.

To calculate numeric item fit statistics, we will use the function `itemfit()` from eRm on the person parameter object (`person.locations.estimate`). This function produces several item fit statistics, including infit mean square error (MSE), outfit MSE, and standardized infit and outfit MSE statistics. We will store the item fit results in a new object called `item.fit`, and then format this object as a dataframe object for easy manipulation and exporting.

```
item.fit.results <- itemfit(person.locations.estimate
   )
item.fit <- cbind.data.frame(item.fit.results$i.
   infitMSQ,
                             item.fit.results$i.
                                outfitMSQ,
                             item.fit.results$i.
                                infitZ,
                             item.fit.results$i.
                                outfitZ)
names(item.fit) <- c("infit_MSE", "outfit_MSE", "std_
   infit", "std_outfit")
```

Next, we will request a summary of the numeric item fit statistics using `summary()`.

```
summary(item.fit)
```

```
##     infit_MSE            outfit_MSE
##   Min.   :0.5296    Min.   :0.4734
##   1st Qu.:0.7671    1st Qu.:0.6360
##   Median :0.8614    Median :0.7457
##   Mean   :0.9968    Mean   :1.0492
##   3rd Qu.:1.0764    3rd Qu.:1.0533
##   Max.   :2.3164    Max.   :3.7723
##     std_infit           std_outfit
##   Min.   :-3.8138   Min.   :-3.2246
##   1st Qu.:-1.6470   1st Qu.:-1.6652
##   Median :-0.9102   Median :-1.0783
```

```
## Mean    :-0.4152   Mean    :-0.1794
## 3rd Qu.: 0.5623    3rd Qu.: 0.3292
## Max.    : 5.9898   Max.    : 8.4184
```

The item.fit object includes mean square error (MSE) and standardized (Z) versions of the outfit and infit statistics for each item included in the analysis. These statistics are summaries of the residuals associated with each item. When data fit Rasch model expectations, the MSE versions of outfit and infit are expected to be close to 1.00 and the standardized versions of outfit and infit are expected to be around 0.00. Please refer to Chapter 3 for a more-detailed discussion of item fit.

Then, we calculate the reliability of separation statistics using the procedure that we described in Chapter 3.

```
## Person separation reliability
person.separation.reliability <- SepRel(person.
    locations.estimate)
person.separation.reliability

##
## Separation Reliability: 0.8874

## Item separation reliability:

# Get Item scores
ItemScores <- colSums(science.responses)

# Get Item SD
ItemSD <- apply(science.responses,2,sd)

# Calculate the se of the Item
ItemSE <- ItemSD/sqrt(length(ItemSD))

# compute the Observed Variance (also known as Total
    Person Variability or Squared Standard Deviation)
SSD.ItemScores <- var(ItemScores)

# compute the Mean Square Measurement error (also
    known as Model Error variance)
Item.MSE <- sum((ItemSE)^2) / length(ItemSE)

# compute the Item Separation Reliability
item.separation.reliability <- (SSD.ItemScores-Item.
    MSE) / SSD.ItemScores
item.separation.reliability

## [1] 0.9999806
```

Person Fit

Next, we will conduct a brief exploration of person fit statistics. To calculate numeric person fit statistics, we will use the function personfit() from *eRm* on the person parameter object (person.locations.estimate). This function produces several person fit statistics, including infit mean square error (MSE), outfit MSE, and standardized infit and outfit MSE statistics. We will store the person fit results in a new object called person.fit, and then format this object as a dataframe for easier manipulation and exporting.

```
person.fit.results <- personfit(person.locations.
   estimate)

person.fit <- cbind.data.frame(person.fit.results$p.
   infitMSQ,
                       person.fit.results$p.
                          outfitMSQ,
                       person.fit.results$p.
                          infitZ,
                       person.fit.results$p.
                          outfitZ)
names(person.fit) <- c("infit_MSE", "outfit_MSE", "
   std_infit", "std_outfit")
```

Next, we will request a summary of the numeric person fit statistics using the summary() function.

```
summary(person.fit)
```

```
##      infit_MSE              outfit_MSE
##   Min.    :0.2754    Min.    :0.2179
##   1st Qu.:0.6635    1st Qu.:0.5470
##   Median :0.8496    Median :0.7221
##   Mean    :0.9588    Mean    :1.0492
##   3rd Qu.:1.0851    3rd Qu.:1.1367
##   Max.    :3.0474    Max.    :4.8182
##     std_infit              std_outfit
##   Min.    :-4.0391    Min.    :-3.1064
##   1st Qu.:-1.2749    1st Qu.:-1.0372
##   Median :-0.3387    Median :-0.4563
##   Mean    :-0.2829    Mean    : 0.0243
##   3rd Qu.: 0.3712    3rd Qu.: 0.5296
##   Max.    : 5.1958    Max.    : 6.6751
```

The *person.fit* object includes mean square error (MSE) and standardized (Z) versions of the outfit and infit statistics for each person. These statistics are summaries of the residuals associated with each person. When data fit Rasch model expectations, the *MSE* versions of outfit and infit are expected to be close to 1.00 and the standardized versions of outfit and infit are expected to be around 0.00. Please refer to Chapter 3 for a more detailed discussion of person fit.

Summarize the Results in Tables

As a final step, we will create tables that summarize the calibrations of the persons, items, and rating scale category thresholds.

Table 1 is an overall model summary table that provides an overview of the logit-scale locations, standard errors, fit statistics, and reliability statistics for items and persons. This type of table is useful for reporting the results from Rasch model analyses because it provides a quick overview of the location estimates and numeric model-data fit statistics for the items and persons in the analysis.

```
RSM_summary.table.statistics <- c("Logit Scale
     Location Mean",
                         "Logit Scale Location
                            SD",
                         "Standard Error Mean",
                         "Standard Error SD",
                         "Outfit MSE Mean",
                         "Outfit MSE SD",
                         "Infit MSE Mean",
                         "Infit MSE SD",
                         "Std. Outfit Mean",
                         "Std. Outfit SD",
                         "Std. Infit Mean",
                         "Std. Infit SD",
                         "Separation.reliability
                            ")

RSM_item.summary.results <- rbind(mean(centered.item.
     locations),
                         sd(centered.item.
                            locations),
                         mean(delta.se),
                         sd(delta.se),
                         mean(item.fit.results$i
                            .outfitMSQ),
                         sd(item.fit.results$i.
                            outfitMSQ),
                         mean(item.fit.results$i
                            .infitMSQ),
```

```
                                                sd(item.fit.results$i.
                                                    infitMSQ),
                                                mean(item.fit.results$i
                                                    .outfitZ),
                                                sd(item.fit.results$i.
                                                    outfitZ),
                                                mean(item.fit.results$i
                                                    .infitZ),
                                                sd(item.fit.results$i.
                                                    infitZ),
                                                item.separation.
                                                    reliability)


RSM_person.summary.results  <-  rbind(mean(person.
    locations$theta),
                                                sd(person.locations$
                                                    theta),
                                                mean(person.locations$
                                                    SE),
                                                sd(person.locations$SE)
                                                    ,
                                                mean(person.fit$outfit_
                                                    MSE),
                                                sd(person.fit$outfit_
                                                    MSE),
                                                mean(person.fit$infit_
                                                    MSE),
                                                sd(person.fit$infit_MSE
                                                    ),
                                                mean(person.fit$std_
                                                    outfit),
                                                sd(person.fit$std_
                                                    outfit),
                                                mean(person.fit$std_
                                                    infit),
                                                sd(person.fit$std_infit
                                                    ),
                                                as.numeric(person.
                                                    separation.
                                                    reliability))

# Round the values for presentation in a table:
RSM_item.summary.results_rounded <- round(RSM_item.
    summary.results, digits = 2)
```

```
RSM_person.summary.results_rounded <- round(RSM_
    person.summary.results[,1], digits = 2)

RSM_Table1 <- cbind.data.frame(RSM_summary.table.
    statistics,
                          RSM_item.summary.results_
                              rounded,
                          RSM_person.summary.results
                              _rounded)

# Add descriptive column labels:
names(RSM_Table1) <- c("Statistic", "Items", "Persons
    ")

# Print the table to the console:
print.data.frame(RSM_Table1, row.names = FALSE)

##                     Statistic Items Persons
##    Logit Scale Location Mean   0.00    1.57
##    Logit Scale Location SD     1.33    1.18
##         Standard Error Mean    0.22    0.38
##         Standard Error SD      0.06    0.10
##            Outfit MSE Mean     1.05    1.05
##            Outfit MSE SD       0.82    0.99
##             Infit MSE Mean     1.00    0.96
##             Infit MSE SD       0.45    0.49
##        Std. Outfit Mean       -0.18    0.02
##        Std. Outfit SD          2.77    1.85
##        Std. Infit Mean        -0.42   -0.28
##        Std. Infit SD           2.35    1.63
##    Separation.reliability      1.00    0.89
```

Table 2 is a table that summarizes the overall calibrations of individual items. For data sets with manageable sample sizes such as the Liking for Science data example in this chapter, we recommend reporting details about each item in a table similar to this one.

```
# Calculate the average rating for each item:
Avg_Rating <- apply(science.responses, 2, mean)

# Combine item calibration results in a table:
RSM_Table2 <- cbind.data.frame(c(1:ncol(science.
    responses)),
                          Avg_Rating,
                          centered.item.locations,
```

```
                              delta.se,
                              item.fit$outfit_MSE,
                              item.fit$std_outfit,
                              item.fit$infit_MSE,
                              item.fit$std_infit)

# Add meaningful column names:
names(RSM_Table2) <- c("Task ID", "Average Rating", "
    Item Location","Item SE","Outfit MSE","Std. Outfit
    ", "Infit MSE","Std. Infit")

# Sort Table 2 by Item difficulty:
RSM_Table2 <- RSM_Table2[order(-RSM_Table2$`Item
    Location`),]

# Round the numeric values (all columns except the
    first one) to 2 digits:
RSM_Table2[, -1] <- round(RSM_Table2[,-1], digits =
    2)

# Print the table to the console:
print.data.frame(RSM_Table2, row.names = FALSE)
```

##	Task ID	Average Rating	Item Location	Item SE
##	5	0.49	2.21	0.18
##	22	0.56	1.98	0.20
##	19	0.67	1.66	0.22
##	4	0.69	1.58	0.22
##	8	0.72	1.50	0.18
##	7	0.92	0.97	0.20
##	9	1.07	0.59	0.25
##	15	1.11	0.49	0.19
##	24	1.13	0.42	0.13
##	3	1.17	0.32	0.20
##	13	1.17	0.32	0.20
##	6	1.21	0.21	0.19
##	16	1.27	0.07	0.45
##	21	1.29	0.00	0.21
##	23	1.43	-0.37	0.18
##	1	1.45	-0.45	0.21
##	14	1.48	-0.53	0.18
##	2	1.55	-0.75	0.18
##	20	1.59	-0.89	0.19
##	11	1.61	-0.98	0.24
##	12	1.69	-1.31	0.18

##	10	1.73	-1.49	0.22
##	18	1.88	-2.45	0.20
##	17	1.93	-3.10	0.35
##	Outfit MSE Std.	Outfit	Infit MSE Std.	Infit
##	3.29	6.77	2.20	5.31
##	3.77	8.42	2.32	5.99
##	1.67	3.16	1.27	1.66
##	0.86	-0.78	0.85	-0.99
##	1.15	0.87	1.06	0.42
##	0.97	-0.15	0.94	-0.42
##	1.13	0.81	1.14	0.99
##	0.90	-0.53	0.92	-0.53
##	0.71	-1.83	0.76	-1.82
##	0.52	-3.22	0.55	-3.81
##	0.72	-1.65	0.79	-1.58
##	0.72	-1.60	0.78	-1.67
##	0.57	-2.56	0.63	-2.92
##	0.72	-1.50	0.80	-1.41
##	0.77	-1.04	0.87	-0.83
##	0.48	-2.68	0.53	-3.65
##	0.62	-1.71	0.76	-1.64
##	0.71	-1.12	0.90	-0.55
##	0.64	-1.35	0.82	-1.03
##	0.47	-2.10	0.60	-2.64
##	0.93	-0.09	1.20	1.02
##	0.56	-1.23	0.77	-1.14
##	1.03	0.25	1.04	0.23
##	1.26	0.56	1.43	1.05

Finally, Table 3 provides a summary of the person calibrations. When there is a relatively large person sample size, it may be more useful to present the results as they relate to individual persons or subsets of the person sample as they are relevant to the purpose of the analysis.

In our person calibration table, we have included the results for all of the children with non-extreme scores. This includes all of the children in our sample except Child #2. For brevity, we only show the first 6 rows of the person calibration table here.

```
# Calculate the average rating for persons who did
    not have extreme scores
Person_Avg_Rating <- apply(person.locations.estimate$
    X.ex,1, mean)

# Combine person calibration results in a table:
```

```
RSM_Table3 <- cbind.data.frame(rownames(person.
    locations),
                                Person_Avg_Rating,
                                person.locations$theta,
                                person.locations$SE,
                                person.fit$outfit_MSE,
                                person.fit$std_outfit,
                                person.fit$infit_MSE,
                                person.fit$std_infit)

# Add meaningful column names:
names(RSM_Table3) <- c("Child ID", "Average Rating",
    "Person Location","Person SE","Outfit MSE","Std.
    Outfit", "Infit MSE","Std. Infit")

# Round the numeric values (all columns except the
    first one) to 2 digits:
RSM_Table3[, -1] <- round(RSM_Table3[,-1], digits =
    2)

# Print the first six rows of the table to the
    console:
print.data.frame(RSM_Table3[1:6,], row.names = FALSE)

##   Child ID Average Rating Person Location
##          1           1.17            1.27
##          3           1.33            1.75
##          4           1.04            0.93
##          5           0.75            0.12
##          6           0.92            0.59
##          7           1.75            3.31
##   Person SE Outfit MSE Std. Outfit Infit MSE
##        0.34       0.83       -0.45      0.93
##        0.35       0.40       -1.99      0.43
##        0.34       0.69       -1.03      0.70
##        0.35       1.20        0.70      0.83
##        0.34       2.36        3.44      1.57
##        0.47       1.12        0.41      1.80
##   Std. Infit
##        -0.20
##        -2.66
##        -1.26
##        -0.58
##         1.99
##         1.81
```

4.3   RSM Analysis with MMLE in *TAM*

The next section of this chapter includes an illustration of RSM analyses with the *Test Analysis Modules* or *TAM* package (Robitzsch et al., 2021) with MMLE. After this illustration, we also demonstrate the use of *TAM* to estimate the RSM with JMLE. These illustrations use the Liking for Science data set that we described earlier in this chapter.

Except where there are significant differences between the *eRm* and *TAM* procedures, we provide fewer details about the analysis procedures and interpretations in this section compared to the first illustration.

Prepare for the Analyses

To get started with the *TAM* package, install and load it into your R environment using the following code.

```
# install.packages("TAM")
library("TAM")
```

To facilitate the example analysis, we will also use the *WrightMap* package (Irribarra and Freund, 2014).

```
# install.packages("WrightMap")
library("WrightMap")
```

If you have not already imported the Liking for Science data and prepared it for analysis as described earlier in this chapter by dropping item 12 and isolating the response matrix, please do so before continuing with the *TAM* analyses.

Run the Rating Scale Model

In order to obtain Rasch-Andrich thresholds (i.e., adjacent-categories thresholds) from our analysis, we need to generate a design matrix for the model that includes specifications for those parameters. We will do this using the designMatrices() function from *TAM* and save the result in a new object called design.matrix.

```
design.matrix <- designMatrices(resp=science.
   responses, modeltype="RSM", constraint = "items")$
   A
```

Now we can run the RSM with our design matrix using the tam.mml() function with several specifications.

```
RSM.science_MMLE <- tam.mml(science.responses,
    irtmodel="RSM", A = design.matrix, constraint = "
    items", verbose = FALSE)
```

Overall Model Summary

After we run the model, we will request a summary of the model results using
the summary() function.

```
summary(RSM.science_MMLE)
```

The summary of the RSM output includes details about the number of iterations, global model fit statistics, a summary of the model parameters, and
several other statistics.

Item Parameters

Next, we will examine the overall item difficulty location and rating scale
category threshold estimates. First, we need to extract the item location
estimates from the model object (RSM.science_MMLE). The item locations are
labeled as *xsi* parameters. We will save the overall item location parameter
estimates and their standard errors for our 24 Liking for Science Items by
selecting the first 24 rows of the *xsi* results in an object called items_MMLE.

```
items_MMLE <- RSM.science_MMLE$xsi[1:(ncol(science.
    responses)),]
```

Next, we need to find the rating scale category threshold parameter estimates
for our rating scale. The Liking for Science rating scale has three categories,
so there are two threshold parameters. The threshold parameters are stored in
a table called item_irt within the RSM object. The first threshold is labeled
tau.Cat1 and the second threshold is labeled tau.Cat2. We can see the values
by printing the table to the console.

```
RSM.science_MMLE$item_irt
```

```
##     item alpha          beta    tau.Cat1  tau.Cat2
## 1    i1     1  -0.449669020  -0.785037  0.785037
## 2    i2     1  -0.749690219  -0.785037  0.785037
## 3    i3     1   0.326379571  -0.785037  0.785037
## 4    i4     1   1.582020472  -0.785037  0.785037
## 5    i5     1   2.199070940  -0.785037  0.785037
## 6    i6     1   0.222036480  -0.785037  0.785037
## 7    i7     1   0.974781105  -0.785037  0.785037
## 8    i8     1   1.507112830  -0.785037  0.785037
## 9    i9     1   0.600174068  -0.785037  0.785037
## 10   i10    1  -1.500765124  -0.785037  0.785037
## 11   i11    1  -0.987962778  -0.785037  0.785037
```

```
## 12   i13      1  -1.313980081  -0.785037  0.785037
## 13   i14      1   0.326379571  -0.785037  0.785037
## 14   i15      1  -0.532200128  -0.785037  0.785037
## 15   i16      1   0.498022740  -0.785037  0.785037
## 16   i17      1   0.080616226  -0.785037  0.785037
## 17   i18      1  -3.126616101  -0.785037  0.785037
## 18   i19      1  -2.473933212  -0.785037  0.785037
## 19   i20      1   1.658247677  -0.785037  0.785037
## 20   i21      1  -0.889691643  -0.785037  0.785037
## 21   i22      1   0.008649293  -0.785037  0.785037
## 22   i23      1   1.979959819  -0.785037  0.785037
## 23   i24      1  -0.369253269  -0.785037  0.785037
## 24   i25      1   0.430310783  -0.785037  0.785037
```

The following code extracts the adjacent-categories threshold parameters from
the `item_irt` table and stores them in an object called `tau.estimates_MMLE`.
Because there is only one set of threshold values for the RSM, we can extract
the threshold estimates for the first item, and the results are applicable to all
of the items in our analysis.

```
# Specify the number of thresholds as the maximum
    observed score in the response matrix (be sure the
    responses begin at category 0):
n.thresholds <- max(science.responses)

## Calculate adjacent-category threshold values:
tau.estimates_MMLE <- NULL

tau.estimates_MMLE <- RSM.science_MMLE$item_irt[1, c
    (4: (4 + (n.thresholds - 1)))]

# Print adjacent-categories threshold estimates to
    the console:
tau.estimates_MMLE
```

```
##     tau.Cat1  tau.Cat2
## 1  -0.785037  0.785037
```

Item Response Functions

Next, we will examine rating scale category probability plots for the items in
our analysis. For brevity, we have only included plots for the first three items
in this book. The specific items to be plotted can be controlled by changing
the items included in the `items.to.plot` object.

```
items.to.plot <- c(1:3)
plot(RSM.science_MMLE, type="items", items = items.to
    .plot)
```

```
## Iteration in WLE/MLE estimation  1   | Maximal
    change  0.882
## Iteration in WLE/MLE estimation  2   | Maximal
    change  0.4971
## Iteration in WLE/MLE estimation  3   | Maximal
    change  0.1235
## Iteration in WLE/MLE estimation  4   | Maximal
    change  9e-04
## Iteration in WLE/MLE estimation  5   | Maximal
    change  0
## ----
##   WLE Reliability= 0.887
```

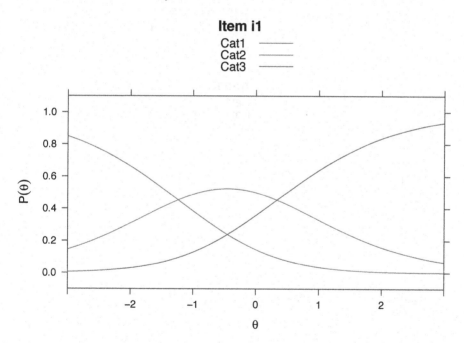

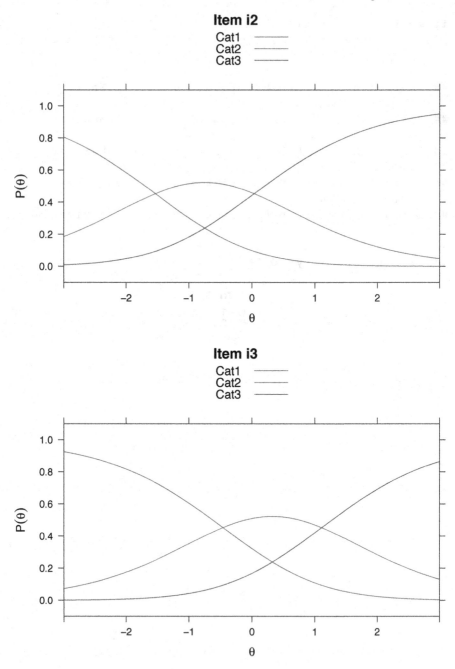

```
##  ..................................................
##  Plots exported in png format into folder:
##  XXXX
```

This code generates plots of rating scale category probabilities for each item. These plots have the same interpretation as the rating scale category probability plots that we generated using *eRm*, where the x-axis is the logit scale that represents the latent variable, the y-axis is the probability for a rating in each category, and individual lines show the conditional probability for a rating in each category.

Item Fit

Next, we will examine numeric item fit indices using the itemfit() function from *TAM*. We will save the results in an object called item.fit_MMLE and then view summary statistics for the fit statistics.

```
MMLE_fit <- tam.fit(RSM.science_MMLE)
item.fit_MMLE <- MMLE_fit$itemfit
summary(item.fit_MMLE)
```

As in the dichotomous Rasch model analysis with TAM (see Chapter 2), the tam.fit() function provides mean square error (MSE) and standardized (t) versions of the Outfit and Infit statistics for Rasch models. The Outfit and Infit statistics are the MSE versions and the Outfit_t and Infit_t statistics are the standardized versions of the statistics. *TAM* also reports a p value for the standardized fit statistics (Outfit_p and Infit_p), along with adjusted significance values (Infit_pholm and Outfit_pholm). Please see Chapter 3 for a detailed consideration of procedures for evaluating item fit, including the use of graphical tools to evaluate item fit.

Person Parameters

Now we will examine person parameter estimates. With the MMLE procedure in *TAM*, person parameters are calculated after the item estimates using the tam.wle() function. The following code calculates person parameter estimates and saves them in an object called person.locations_MMLE.

```
# Use the tam.wle function to calculate person
    location parameters:
person.locations.estimate_MMLE <- tam.wle(RSM.science
    _MMLE)

## Iteration in WLE/MLE estimation 1   | Maximal
    change  0.882
## Iteration in WLE/MLE estimation 2   | Maximal
    change  0.4971
## Iteration in WLE/MLE estimation 3   | Maximal
    change  0.1235
## Iteration in WLE/MLE estimation 4   | Maximal
    change  9e-04
## Iteration in WLE/MLE estimation 5   | Maximal
    change  0
```

```
## ----
##   WLE Reliability= 0.887
```

```
# Store person parameters and their standard errors
    in a dataframe object:
person.locations_MMLE <- cbind.data.frame(person.
    locations.estimate_MMLE$theta,
                                    person.locations
                                          .estimate_
                                          MMLE$error)
```

```
names(person.locations_MMLE) <- c("theta", "SE")
```

```
# View summary statistics for person parameters:
summary(person.locations_MMLE)
```

```
##       theta                SE
##   Min.   :-1.5815   Min.    :0.3365
##   1st Qu.: 0.0373   1st Qu.:0.3379
##   Median : 0.6006   Median :0.3472
##   Mean   : 0.8280   Mean    :0.3921
##   3rd Qu.: 1.4708   3rd Qu.:0.3802
##   Max.   : 5.3850   Max.    :1.4721
```

Person Fit

We can evaluate person fit using the `tam.personfit()` function from TAM. This function uses the model object as an argument and it produces infit and outfit statistics, as well as standardized versions of these statistics, for each person in the response matrix.

```
person.fit.results_MMLE <- tam.personfit(RSM.science_
    MMLE)
summary(person.fit.results_MMLE)
```

```
##   outfitPerson        outfitPerson_t
##   Min.   :0.02029   Min.    :-3.08806
##   1st Qu.:0.52565   1st Qu.:-1.04125
##   Median :0.71316   Median :-0.45831
##   Mean   :0.97920   Mean    :-0.04494
##   3rd Qu.:1.12658   3rd Qu.: 0.50305
##   Max.   :4.08127   Max.    : 5.28727
##    infitPerson        infitPerson_t
##   Min.   :0.04679   Min.    :-4.0274
##   1st Qu.:0.65236   1st Qu.:-1.2797
```

```
##    Median  :0.83053    Median  :-0.4127
##    Mean    :0.93502    Mean    :-0.3221
##    3rd Qu.:1.04901     3rd Qu.: 0.2676
##    Max.    :3.04482    Max.    : 5.1969
```

Wright Map

Next, we will create a Wright Map from our model results. This display will provide us with an overview of the distribution of person, item, and threshold parameters. We will create the plot using the *WrightMap* package function IRT.WrightMap() on the model object (RSM.science_MMLE). The following code prepares the parameter estimates and uses them to plot the Wright Map.

```
# Combine item estimates with thresholds:
n.items <- ncol(science.responses)

thresholds_MMLE <- matrix(data = NA, nrow = n.items,
    ncol = n.thresholds)

tau.estimates_MMLE <- as.vector(as.numeric(tau.
    estimates_MMLE))

for(i in 1:n.thresholds){
  items.thresholds <- items_MMLE + tau.estimates_MMLE
    [i]
  thresholds_MMLE[, i] <- items.thresholds[1:n.items,
    1]
}


# Plot the Wright Map
wrightMap(thetas = person.locations_MMLE$theta,
        thresholds = thresholds_MMLE,
        main.title = "Liking for Science Rating
            Scale Model Wright Map",
        show.thr.lab   = TRUE, dim.names = "",
        label.items.rows= 2)
```

Liking for Science Rating Scale Model Wright Map

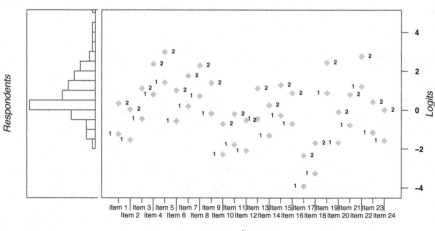

```
##              [,1]          [,2]
##    [1,]  -1.2347060    0.33536797
##    [2,]  -1.5347272    0.03534678
##    [3,]  -0.4586574    1.11141657
##    [4,]   0.7969835    2.36705747
##    [5,]   1.4140339    2.98410793
##    [6,]  -0.5630005    1.00707347
##    [7,]   0.1897441    1.75981810
##    [8,]   0.7220758    2.29214982
##    [9,]  -0.1848629    1.38521106
##   [10,]  -2.2858021   -0.71572813
##   [11,]  -1.7729998   -0.20292578
##   [12,]  -2.0990171   -0.52894309
##   [13,]  -0.4586574    1.11141657
##   [14,]  -1.3172371    0.25283687
##   [15,]  -0.2870143    1.28305973
##   [16,]  -0.7044208    0.86565322
##   [17,]  -3.9116531   -2.34157911
##   [18,]  -3.2589702   -1.68889622
##   [19,]   0.8732107    2.44328467
##   [20,]  -1.6747286   -0.10465465
##   [21,]  -0.7763877    0.79368629
##   [22,]   1.1949228    2.76499681
##   [23,]  -1.1542903    0.41578373
##   [24,]  -1.5700740    0.00000000
```

In this *Wright Map* display, the results from the RSM analysis of the Liking for Science data are summarized graphically. The figure is organized as follows:

The left panel of the plot shows a histogram of respondent (children) locations on the logit scale that represents the latent variable. Units on the logit scale are shown on the far-right axis of the plot (labeled *Logits*).

The large central panel of the plot shows the rating scale category threshold estimates specific to each item on the logit scale that represents the latent variable. Light gray diamond shapes show the logit-scale location of the threshold estimates for each item, as labeled on the x-axis. Thresholds are labeled using tau an integer that shows the threshold number. In our example, τ_1 is the threshold between rating scale categories $x = 0$ and $x = 1$, and τ_2 is the threshold between rating scale categories $x = 1$ and $x = 2$.

Even though it is not appropriate to fully interpret item and person locations on the logit scale until there is evidence of acceptable model-data fit, we recommend examining the Wright Map during the preliminary stages of an item analysis to get a general sense of the model results and to identify any potential scoring or data entry errors.

A quick glance at the Wright Map suggests that, on average, the persons are located higher on the logit scale compared to the average item threshold locations. In addition, there appears to be a relatively wide spread of person and item locations on the logit scale, such that the Liking for Science questionnaire appears to be a useful tool for identifying differences in children's attitudes toward science activities as well as the difficulty to find each of the activities as favorable.

4.4   RSM Analysis with JMLE in TAM

In the following section, we provide an illustration of RSM analyses with the *Test Analysis Modules* or *TAM* package (Robitzsch et al., 2021) with JMLE. Except where there are significant differences between the MMLE and JMLE procedures, we provide fewer details about the analysis procedures and interpretations compared to the *eRm* and *TAM* MMLE illustrations.

Prepare for the Analyses

Before beginning the JMLE RSM analysis, please ensure that you have installed the *TAM* and *WrightMap* packages and loaded them into your working environment. In addition, please ensure that you have imported the Liking for Science data, removed item 12, and isolated the response matrix as described earlier in the chapter.

Run the Rating Scale Model

We will use the `tam.jml()` function to run the RSM with JMLE and store the results in an object called `RSM.science_JMLE`. As with the MMLE procedure, in order to obtain Rasch-Andrich thresholds (i.e., adjacent-categories thresholds), we need to generate a design matrix for the model that includes specifications for those parameters. We will do this using the `designMatrices()` function from *TAM* and save the result in a new object called `design.matrix`.

```
design.matrix <- designMatrices(resp=science.
    responses, modeltype="RSM", constraint = "items")$
    A
```

Now we can run the RSM with our design matrix. We have also added a few other specifications to maximize the comparability of the interpretation of the results from this analysis with results from the Winsteps (2020b) and Facets (2020a) software programs.

```
RSM.science_JMLE <- tam.jml(science.responses, A =
    design.matrix, constraint = "items",control=list(
    maxiter=500), version=2, verbose = FALSE)
```

Overall Model Summary

After we run the model, we will request a summary of the model results using the `summary()` function.

```
summary(RSM.science_JMLE)
```

The summary of the RSM output includes details about the number of iterations, global model fit statistics, a summary of the model parameters, and several other statistics.

Item Parameters

Next, we will examine the item difficulty location and rating scale category threshold estimates. First, we need to extract the item location estimates from the model object (`RSM.science_MMLE`). The item locations are stored in the `item` table within the model object. The item locations are stored in the `xsi.item` column of this table. First, we will view the item table by printing it to the console. Then, we will save the overall item location parameter estimates in an object called `items_MMLE`.

```
RSM.science_JMLE$item
```

```
##      item  N          M      xsi.item AXsi_.Cat1
## i1     i1 75 1.4533333 -0.4574934878 -1.2681805
## i2     i2 75 1.5466667 -0.7552041947 -1.5658912
```

```
## i3      i3 75 1.1733333  0.3170099599 -0.4936770
## i4      i4 75 0.6933333  1.5859579282  0.7752710
## i5      i5 75 0.4933333  2.2217139684  1.4110270
## i6      i6 75 1.2133333  0.2124921916 -0.5981948
## i7      i7 75 0.9200000  0.9689816136  0.1582946
## i8      i8 75 0.7200000  1.5095479845  0.6988610
## i9      i9 75 1.0666667  0.5913466249 -0.2193404
## i10    i10 75 1.7333333 -1.4968099225 -2.3074969
## i11    i11 75 1.6133333 -0.9910288297 -1.8017158
## i13    i13 75 1.6933333 -1.3127959249 -2.1234829
## i14    i14 75 1.1733333  0.3170099599 -0.4936770
## i15    i15 75 1.4800000 -0.5394778715 -1.3501648
## i16    i16 75 1.1066667  0.4886897578 -0.3219972
## i17    i17 75 1.2666667  0.0710178953 -0.7396691
## i18    i18 75 1.9333333 -3.0878980651 -3.8985850
## i19    i19 75 1.8800000 -2.4504761670 -3.2611631
## i20    i20 75 0.6666667  1.6640135967  0.8533266
## i21    i21 75 1.5866667 -0.8938071367 -1.7044941
## i22    i22 75 1.2933333 -0.0008840094 -0.8115710
## i23    i23 75 0.5600000  1.9947495520  1.1840626
## i24    i24 75 1.4266667 -0.3775422823 -1.1882293
## i25    i25 75 1.1333333  0.4208868590 -0.3898001
##          AXsi_.Cat2 B.Cat1.Dim1 B.Cat2.Dim1
## i1    -0.914986976           1           2
## i2    -1.510408389           1           2
## i3     0.634019920           1           2
## i4     3.171915856           1           2
## i5     4.443427937           1           2
## i6     0.424984383           1           2
## i7     1.937963227           1           2
## i8     3.019095969           1           2
## i9     1.182693250           1           2
## i10   -2.993619845           1           2
## i11   -1.982057659           1           2
## i13   -2.625591850           1           2
## i14    0.634019920           1           2
## i15   -1.078955743           1           2
## i16    0.977379516           1           2
## i17    0.142035791           1           2
## i18   -6.175796130           1           2
## i19   -4.900952334           1           2
## i20    3.328027193           1           2
## i21   -1.787614273           1           2
## i22   -0.001768019           1           2
## i23    3.989499104           1           2
```

```
## i24  -0.755084565              1              2
## i25   0.841773718              1              2
```

```
items_JMLE <- RSM.science_JMLE$item$xsi.item
```

Next, we need to find the adjacent-categories rating scale category threshold
parameter estimates for our rating scale. In our example with the Liking for
Science data, our rating scale has three categories, so there are two threshold
parameters. With the JMLE function in *TAM*, adjacent-categories threshold
values are stored in the item1 table within the model object. These threshold
estimates are the final values in the xsi column.

```
RSM.science_JMLE$item1
```

##	xsi.label	xsi.index	xsi	se.xsi
## 1	i1	1	-0.4574934878	0.13927663
## 2	i2	2	-0.7552041947	0.14256938
## 3	i3	3	0.3170099599	0.13420889
## 4	i4	4	1.5859579282	0.13776306
## 5	i5	5	2.2217139684	0.14419747
## 6	i6	6	0.2124921916	0.13457283
## 7	i7	7	0.9689816136	0.13425493
## 8	i8	8	1.5095479845	0.13716421
## 9	i9	9	0.5913466249	0.13374407
## 10	i10	10	-1.4968099225	0.15303171
## 11	i11	11	-0.9910288297	0.14560813
## 12	i13	12	-1.3127959249	0.15022452
## 13	i14	13	0.3170099599	0.13420889
## 14	i15	14	-0.5394778715	0.14011662
## 15	i16	15	0.4886897578	0.13383474
## 16	i17	16	0.0710178953	0.13522854
## 17	i18	17	-3.0878980651	0.17561557
## 18	i19	18	-2.4504761670	0.16762246
## 19	i20	19	1.6640135967	0.13841808
## 20	i21	20	-0.8938071367	0.14431437
## 21	i22	21	-0.0008840094	0.13563297
## 22	i23	22	1.9947495520	0.14163552
## 23	i24	23	-0.3775422823	0.13850953
## 24	Cat1	24	-0.8106869777	0.05342681

With the JMLE function in *TAM*, the highest threshold parameter is not
reported and must be calculated manually. In our example, this is the threshold
between category $x = 1$ and $x = 2$, and it is the second threshold. The following
code calculates the adjacent-categories threshold parameters and stores them
in an object called tau.estimates_JMLE:

```
# Specify the number of thresholds as the maximum
    observed score in the response matrix (be sure the
    responses begin at category 0):
n.thresholds <- max(science.responses)

## Calculate adjacent-category threshold values:
tau.estimates_JMLE <- NULL

# Find all but the final threshold estimate:
for(tau in 1: (n.thresholds - 1)){
    tau.estimates_JMLE[tau] <- RSM.science_JMLE$item1$
        xsi[(ncol(science.responses) - 1) + tau ]
}

# Calculate the final threshold estimate:
tau.estimates_JMLE[n.thresholds] <- -(sum(tau.
    estimates_JMLE))

# Print adjacent-categories threshold estimates to
    the console:
tau.estimates_JMLE

## [1] -0.810687   0.810687
```

Item Response Functions

Next, we will examine rating scale category probability plots for the items in our analysis. As before, we only create plots for the first three items here.

```
items.to.plot <- c(1:3)
plot(RSM.science_JMLE, type="items", items = items.to
    .plot)
```

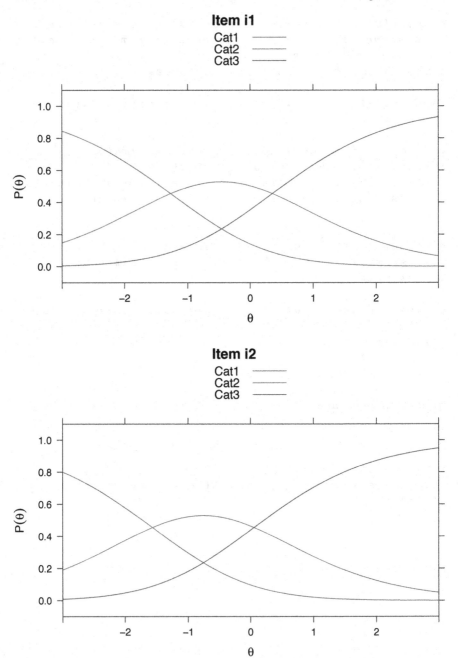

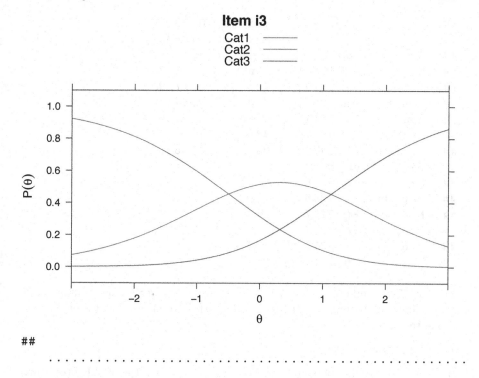

Item i3

Cat1 ———
Cat2 ———
Cat3 ———

```
##
```

....

```
## Plots exported in png format into folder:
## XXXX
```

This code generates plots of rating scale category probabilities for each item. These plots have the same interpretation as the rating scale category probability plots presented earlier in this chapter.

Item Fit

Next, we will examine numeric item fit indices using the `tam.fit()` function. We will save the results in an object called `item.fit_JMLE` and then view summary statistics for the fit statistics.

```
JMLE_fit <- tam.fit(RSM.science_JMLE)

item.fit_JMLE <- JMLE_fit$fit.item
summary(item.fit_JMLE)

##     item             outfitItem
## Length:24         Min.   :0.4685
## Class :character  1st Qu.:0.6304
## Mode  :character  Median :0.7463
##                   Mean   :0.9979
```

```
##                        3rd  Qu.:0.9335
##                        Max.    :3.3232
##    outfitItem_t           infitItem
##   Min.     :-2.43501   Min.     :0.5388
##   1st  Qu.:-1.20199   1st  Qu.:0.7729
##   Median :-0.73631    Median :0.8741
##   Mean     : 0.01117   Mean     :1.0083
##   3rd  Qu.: 0.08614   3rd  Qu.:1.0976
##   Max.     : 7.19795   Max.     :2.3572
##    infitItem_t
##   Min.     :-3.6978
##   1st  Qu.:-1.5345
##   Median :-0.8157
##   Mean     :-0.3222
##   3rd  Qu.: 0.6672
##   Max.     : 6.1617
```

The `tam.fit()` function provides mean square error (MSE) and standardized
(t) versions of the outfit and infit statistics for Rasch models for each item
estimate.

Person Parameters

Now we will examine person parameter estimates. With the JMLE procedure
in *TAM*, person parameters are calculated at the same time as the item
parameters. This means that we can extract the person parameter estimates
from the model object without any additional iterations. The following code
calculates person parameter estimates and saves them in an object called
person.locations_JMLE.

```
# Store person parameters and their standard errors
    in a dataframe object:
person.locations_JMLE <- cbind.data.frame(RSM.science
   _JMLE$theta,

                                  RSM.science_JMLE
                                       $errorWLE)
# Add meaningful column names:
names(person.locations_JMLE) <- c("theta", "SE")

# View summary statistics for person parameters:
summary(person.locations_JMLE)

##        theta                 SE
##   Min.    :-1.66163   Min.     :0.3420
##   1st Qu.: 0.03455   1st Qu.:0.3435
##   Median : 0.62466   Median :0.3530
##   Mean    : 0.87738   Mean     :0.3940
```

```
## 3rd Qu.: 1.53961    3rd Qu.:0.3859
## Max.   : 5.97167    Max.   :1.1795
```

Person Fit

Next, we will examine numeric person fit indices using the `tam.fit()` function. We will save the results in an object called `person.fit_JMLE` and then view summary statistics for the fit statistics.

```
JMLE_fit <- tam.fit(RSM.science_JMLE)

person.fit_JMLE <- JMLE_fit$fit.person
summary(person.fit_JMLE)

##    outfitPerson        outfitPerson_t
## Min.    :0.01164    Min.    :-3.1237
## 1st Qu.:0.54991    1st Qu.:-1.0090
## Median :0.73012    Median :-0.3986
## Mean    :0.99788    Mean    : 0.0141
## 3rd Qu.:1.15647    3rd Qu.: 0.6122
## Max.    :4.12918    Max.    : 5.3910
##    infitPerson         infitPerson_t
## Min.    :0.02743    Min.    :-4.0104
## 1st Qu.:0.67191    1st Qu.:-1.2205
## Median :0.89781    Median :-0.2832
## Mean    :0.96690    Mean    :-0.2184
## 3rd Qu.:1.11210    3rd Qu.: 0.4551
## Max.    :3.07743    Max.    : 5.2383
```

The `tam.fit()` function provides mean square error (MSE) and standardized (t) versions of the Outfit and Infit statistics for Rasch models for each person location estimate.

Wright Map

Finally, we will create a Wright Map from our model results. This display will provide us with an overview of the distribution of person, item, and threshold parameters. We will create the plot using the *WrightMap* package function `IRT.WrightMap` on the model object (`RSM.science_JMLE`). The following code prepares the parameter estimates and plots the Wright Map using them.

```
# Combine item estimates with thresholds:
n.items <- ncol(science.responses)

thresholds_JMLE <- matrix(data = NA, nrow = n.items,
    ncol = n.thresholds)
```

```
tau.estimates_JMLE <- as.vector(as.numeric(tau.
    estimates_JMLE))

for(i in 1:n.thresholds){
  items.thresholds <- items_JMLE + tau.estimates_JMLE
      [i]
  thresholds_JMLE[, i] <- items.thresholds
}

# Plot the Wright Map
wrightMap(thetas = person.locations_JMLE$theta,
        thresholds = thresholds_JMLE,
        main.title = "Liking for Science Rating
            Scale Model Wright Map: JMLE",
        show.thr.lab   = TRUE, dim.names = "",
        label.items.rows= 2)
```

Liking for Science Rating Scale Model Wright Map: JMLE

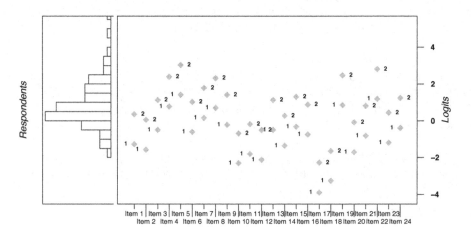

```
##                    [,1]            [,2]
## [1,]  -1.2681805   0.35319349
## [2,]  -1.5658912   0.05548278
## [3,]  -0.4936770   1.12769694
## [4,]   0.7752710   2.39664491
## [5,]   1.4110270   3.03240095
## [6,]  -0.5981948   1.02317917
## [7,]   0.1582946   1.77966859
## [8,]   0.6988610   2.32023496
## [9,]  -0.2193404   1.40203360
```

```
## [10,]  -2.3074969  -0.68612294
## [11,]  -1.8017158  -0.18034185
## [12,]  -2.1234829  -0.50210895
## [13,]  -0.4936770   1.12769694
## [14,]  -1.3501648   0.27120911
## [15,]  -0.3219972   1.29937674
## [16,]  -0.7396691   0.88170487
## [17,]  -3.8985850  -2.27721109
## [18,]  -3.2611631  -1.63978919
## [19,]   0.8533266   2.47470057
## [20,]  -1.7044941  -0.08312016
## [21,]  -0.8115710   0.80980297
## [22,]   1.1840626   2.80543653
## [23,]  -1.1882293   0.43314470
## [24,]  -0.3898001   1.23157384
```

In this *Wright Map* display, the results from the RSM analysis of the Liking for Science data are summarized graphically. The figure is organized in the same way as we described for the MMLE RSM analysis earlier in this chapter.

4.5 Example Results Section

Table 4.1 presents a summary of the results from the analysis of the Liking for Science data (Wright and Masters, 1982) using the Rating Scale Model (RSM) (Andrich, 1978). One item was excluded (item 12) because the responses to this item did not include any observations in the lowest rating scale category ($x = 0$). In addition, one child was excluded from the analysis (Child #2), who gave extreme responses to all of the items ($x = 2$).

```
# Print Table 1:
tab1 <- knitr::kable(
  RSM_Table1, booktabs = TRUE,
  caption = 'Model Summary Table'
)
tab1 %>%
  kable_styling(latex_options = "scale_down")
```

Specifically, Table 4.1 summarizes the calibration of the children with non-extreme responses ($N = 74$) and the items ($N = 24$) using average logit-scale calibrations, standard errors, and model-data fit statistics. Examination of the results indicates that, on average, the students were located higher on the logit scale ($M = 1.57$, $SD = 1.18$), compared to items, whose locations were

TABLE 4.1
Model Summary Table

Statistic	Items	Persons
Logit Scale Location Mean	0.00	1.57
Logit Scale Location SD	1.33	1.18
Standard Error Mean	0.22	0.38
Standard Error SD	0.06	0.10
Outfit MSE Mean	1.05	1.05
Outfit MSE SD	0.82	0.99
Infit MSE Mean	1.00	0.96
Infit MSE SD	0.45	0.49
Std. Outfit Mean	−0.18	0.02
Std. Outfit SD	2.77	1.85
Std. Infit Mean	−0.42	−0.28
Std. Infit SD	2.35	1.63
Separation.reliability	1.00	0.89

centered at zero logits ($M = 0.00$, $SD = 1.33$). This finding suggests that the items were relatively easy to endorse among the sample of children who participated in this study. The average values of the Standard Error (SE) were comparable for children ($M = 0.38$) and items ($M = 0.22$). Average values of model-data fit statistics indicate overall adequate fit to the model, with average outfit and infit mean square error (MSE) statistics around 1.00, and average standardized outfit and infit statistics near the expected value of 0.00 when data fit the model. However, there was some variability in item fit and person fit, as indicated by relatively large standard errors for the fit statistics. Additional investigation into item fit and person fit is warranted.

```
# Print Table 2:
tab2 <- knitr::kable(
  RSM_Table2, booktabs = TRUE,
  caption = 'Item Calibration'
)
tab2 %>%
  kable_styling(latex_options = "scale_down")
```

Table 4.2 includes detailed results for the 24 items included in the analysis, where items are ordered by their overall logit-scale location (i.e., item difficulty) from high (difficult to endorse) to low (easy to endorse). For each item, the average rating is presented, followed by the overall logit-scale location (δ), SE, and model-data fit statistics. Examination of these results indicates that Item 5 was the most difficult to endorse ($AverageRating = 0.49$; $\delta = 2.21$;

TABLE 4.2
Item Calibration

	Task ID	Average Rating	Item Location	Item SE	Outfit MSE	Std. Outfit	Infit MSE	Std. Infit
i5	5	0.49	2.21	0.18	3.29	6.77	2.20	5.31
i23	22	0.56	1.98	0.20	3.77	8.42	2.32	5.99
i20	19	0.67	1.66	0.22	1.67	3.16	1.27	1.66
i4	4	0.69	1.58	0.22	0.86	−0.78	0.85	−0.99
i8	8	0.72	1.50	0.18	1.15	0.87	1.06	0.42
i7	7	0.92	0.97	0.20	0.97	−0.15	0.94	−0.42
i9	9	1.07	0.59	0.25	1.13	0.81	1.14	0.99
i16	15	1.11	0.49	0.19	0.90	−0.53	0.92	−0.53
i25	24	1.13	0.42	0.13	0.71	−1.83	0.76	−1.82
i3	3	1.17	0.32	0.20	0.52	−3.22	0.55	−3.81
i14	13	1.17	0.32	0.20	0.72	−1.65	0.79	−1.58
i6	6	1.21	0.21	0.19	0.72	−1.60	0.78	−1.67
i17	16	1.27	0.07	0.45	0.57	−2.56	0.63	−2.92
i22	21	1.29	0.00	0.21	0.72	−1.50	0.80	−1.41
i24	23	1.43	−0.37	0.18	0.77	−1.04	0.87	−0.83
i1	1	1.45	−0.45	0.21	0.48	−2.68	0.53	−3.65
i15	14	1.48	−0.53	0.18	0.62	−1.71	0.76	−1.64
i2	2	1.55	−0.75	0.18	0.71	−1.12	0.90	−0.55
i21	20	1.59	−0.89	0.19	0.64	−1.35	0.82	−1.03
i11	11	1.61	−0.98	0.24	0.47	−2.10	0.60	−2.64
i13	12	1.69	−1.31	0.18	0.93	−0.09	1.20	1.02
i10	10	1.73	−1.49	0.22	0.56	−1.23	0.77	−1.14
i19	18	1.88	−2.45	0.20	1.03	0.25	1.04	0.23
i18	17	1.93	−3.10	0.35	1.26	0.56	1.43	1.05

$SE = 0.18$), followed by Item 23 (*AverageRating* $= 0.56$; $\delta = 1.98$; $SE = 0.20$). The easiest item to endorse was Item 18 (*AverageRating* $= 1.93$; $\delta = -3.10$; $SE = 0.35$).

```
# Print Table 3:
tab3 <- knitr::kable(
  head(RSM_Table3,10), booktabs = TRUE,
  caption = 'Person Calibration'
)
tab3 %>%
  kable_styling(latex_options = "scale_down")
```

TABLE 4.3

Person Calibration

Child ID	Average Rating	Person Location	Person SE	Outfit MSE	Std. Outfit	Infit MSE	Std. Infit
1	1.17	1.27	0.34	0.83	−0.45	0.93	−0.20
3	1.33	1.75	0.35	0.40	−1.99	0.43	−2.66
4	1.04	0.93	0.34	0.69	−1.03	0.70	−1.26
5	0.75	0.12	0.35	1.20	0.70	0.83	−0.58
6	0.92	0.59	0.34	2.36	3.44	1.57	1.99
7	1.75	3.31	0.47	1.12	0.41	1.80	1.81
8	1.29	1.62	0.35	0.83	−0.39	0.98	0.03
9	0.92	0.59	0.34	1.80	2.28	1.37	1.37
10	1.08	1.04	0.34	0.65	−1.19	0.70	−1.25
11	1.38	1.87	0.36	0.79	−0.43	0.97	−0.01

Table 4.3 includes detailed results for first 10 children who participated in the Liking for Science survey. For each child, the average rating is presented, followed by their logit-scale location estimate (θ), *SE*, and model-data fit statistics.

```
plotPImap(RSM.science, main = "Liking for Science
    Rating Scale Model Wright Map", sorted = TRUE,
    irug = FALSE)
```

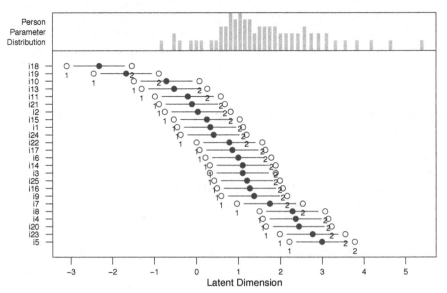

Liking for Science Rating Scale Model Wright Map

Figure 1 illustrates the calibrations of the children and items on the logit scale that represents the latent variable. The calibrations shown in this figure correspond to the results presented in Tables 4.2 and 4.3 for items and children, respectively.Starting at the bottom of the figure, the horizontal axis (labeled *Latent Dimension*) is the logit scale that represents the latent variable. Lower numbers on this scale indicate less-favorable attitudes toward science activities, and higher numbers indicate more-favorable attitudes toward science activities. The central panel of the figure shows item difficulty locations on the logit scale for the 24 Liking for Science items included in the analysis; the y-axis for this panel shows the item labels. The items are ordered according to their difficulty order, as estimated with the RSM. Easier-to-endorse items appear at the top of the figure, and harder-to-endorse items appear at the bottom of the figure; item labels are shown on the y-axis. For each item, a solid circle plotting symbol shows the overall location estimate. This solid circle symbol is connected to two open-circle symbols that show the locations of the rating scale category thresholds. Each threshold is labeled with a 1 to indicate the threshold between rating scale category $x = 0$ and $x = 1$, or a 2 to indicate the threshold between rating scale category $x = 1$ and $x = 2$. Examination of the item locations reveals that the threshold locations were ordered as expected, with τ_1 located lower on the logit scale ($\tau_1 = -0.78$) compared to the location estimate for τ_2 ($\tau_2 = 0.78$). The upper panel of the figure shows a histogram of person (in this case, children) location estimates on the logit scale.

4.6 Exercise

Use the *eRm* or *TAM* package to estimate item, threshold, and person locations with the RSM for the Exercise 4 data. The Exercise 4 data include responses from 500 participants to an attitude survey that includes 20 items with a 5-category rating scale. Then, try writing a results section similar to the example in this chapter to report your findings.

5

Partial Credit Model

This chapter provides a basic overview of the Partial Credit Model (PCM) (Masters, 1982), along with guidance for analyzing data with the PCM using R. We use the same example data set from Chapter 4 of this book that includes participant responses to an attitude survey about science activities to illustrate the analysis using Conditional Maximum Likelihood Estimation (CMLE) via the *eRm* package (Mair et al., 2021). We also demonstrate PCM analyses using Marginal Maximum Likelihood Estimation (MMLE) and Joint Maximum Likelihood Estimation (JMLE) via the *TAM* package (Robitzsch et al., 2021). After the analyses are complete, we present an example description of the results. The chapter concludes with a challenge exercise.

Overview of the Partial Credit Model

Masters (1982) proposed the PCM for use with ordinal item responses that are scored in more than two categories (e.g., data from attitude scales or performance assessments). Similar to the Rating Scale Model (RSM; Andrich (1978); see Chapter 4), the PCM provides estimates of *person locations, item locations,* and *rating scale category thresholds* on a log-odds scale that represents the latent variable. However, whereas the RSM provides one set of rating scale category thresholds that are estimated using all item responses, the PCM provides separate rating scale category thresholds for each item included in the analysis. Item-specific thresholds are useful in many practical situations, including instruments that include multiple scale lengths and cases where some rating scale categories are not observed in responses to one or more items. In addition, the PCM is useful in contexts where it is important to verify comparable rating scale functioning across items, and to evaluate the measurement quality of rating scales specific to individual items.

The PCM can be stated in log-odds form as follows:

$$ ln\left[\frac{P_{n_i(xi=k)}}{P_{n_i(xi=k-1)}}\right] = \theta_n - \delta_i - \tau_{ik} \qquad (5.1) $$

In the PCM, θ is the person's ability, δ is the item's difficulty, and τ_{ik} is the rating scale category threshold specific to item i. As in the RSM, the threshold is the location on the logit scale at which there is an equal probability for a rating in category k and category $k - 1$. For a rating scale made up of m

DOI: 10.1201/9781003174660-5

categories, there are $m - 1$ rating scale category thresholds. Thresholds (τ_k) are estimated separately for each element of a selected facet, such as items. They are not necessarily evenly spaced or ordered as expected.

Rasch Model Requirements

The PCM is based on the same requirements of unidimensionality, local independence, and invariance that we discussed in Chapter 2 for the dichotomous Rasch model. In practice, researchers should evaluate item responses for evidence that they approximate Rasch model requirements before examining model estimates in detail. Chapter 3 included details about model-data fit analysis procedures that can also be applied to the PCM. In the current chapter, we provide code for calculating some popular residual-based fit indices for items and persons based on the PCM.

5.1 Example Data: Liking for Science

The example data set for this chapter is the same data set that we used in Chapter 4 of this book. The data include a group of 75 children's responses to the 25-item *Liking for Science* questionnaire, which was designed to measure their attitudes toward science activities. The data were originally published in Wright and Masters (1982). Each item stem included a science activity, and three response options: $0 = Dislike$, $1 = Not\ Sure/Don't\ Care$, and $2 = Like$, such that responses in higher categories indicated more-favorable attitudes toward science activities.

5.2 PCM Analysis with CMLE in *eRm*

In the next section, we provide a step-by-step demonstration of a PCM analysis with CMLE using the *eRm* package. We encourage readers to use the example data set that is provided in the online supplement to conduct the analysis along with us.

Prepare for the Analyses

We will use the *eRm* package (Mair et al., 2021) as the first package with which we demonstrate PCM analyses. We selected *eRm* for the first illustrations in the current chapter because it includes functions for applying the PCM that are relatively straightforward to use and interpret. Please note that the *eRm* package uses the CMLE method to estimate Rasch model parameters.

As a result, estimates from the *eRm* package are not directly comparable to estimates obtained using other estimation methods. At the end of this chapter, we have included an illustration of RSM analyses with the *TAM* package (Robitzsch et al., 2021) with MMLE. We also provide an illustration with *TAM* using JMLE, which produces comparable estimates to some popular standalone Rasch software programs, such as Winsteps (Winsteps, 2020b) and Facets (Facets, 2020a).

```
#install.packages("eRm")
library("eRm")
```

Now that we have installed and loaded the package to our R session, we are ready to import the data. We will use the function **read.csv()** to import the comma-separated values (.csv) file that contains the Liking for Science survey data. We encourage readers to use their preferred method for importing data files into R or R Studio.

Please note that if you use **read.csv()** to import the data, you will first need to specify the file path to the location at which the data file is stored on your computer or set your working directory to the folder in which you have saved the data.

We will import the data using **read.csv()** and store it in an object called **science**.

```
science <- read.csv("liking_for_science.csv")
```

Next, we will explore the data using descriptive statistics with the **summary()** function.

```
summary(science)
```

```
##      student              i1               i2
##   Min.   : 1.0     Min.   :0.000    Min.   :0.000
##   1st Qu.:19.5     1st Qu.:1.000    1st Qu.:1.000
##   Median :38.0     Median :1.000    Median :2.000
##   Mean   :38.0     Mean   :1.453    Mean   :1.547
##   3rd Qu.:56.5     3rd Qu.:2.000    3rd Qu.:2.000
##   Max.   :75.0     Max.   :2.000    Max.   :2.000
##        i3               i4
##   Min.   :0.000    Min.   :0.0000
##   1st Qu.:1.000    1st Qu.:0.0000
##   Median :1.000    Median :1.0000
##   Mean   :1.173    Mean   :0.6933
##   3rd Qu.:2.000    3rd Qu.:1.0000
##   Max.   :2.000    Max.   :2.0000
##        i5               i6               i7
##   Min.   :0.0000   Min.   :0.000    Min.   :0.00
```

```
## 1st Qu.:0.0000    1st Qu.:1.000    1st Qu.:0.00
## Median :0.0000    Median :1.000    Median :1.00
## Mean   :0.4933    Mean   :1.213    Mean   :0.92
## 3rd Qu.:1.0000    3rd Qu.:2.000    3rd Qu.:1.50
## Max.   :2.0000    Max.   :2.000    Max.   :2.00
##       i8               i9              i10
## Min.   :0.00     Min.   :0.000    Min.   :0.000
## 1st Qu.:0.00     1st Qu.:0.000    1st Qu.:2.000
## Median :1.00     Median :1.000    Median :2.000
## Mean   :0.72     Mean   :1.067    Mean   :1.733
## 3rd Qu.:1.00     3rd Qu.:2.000    3rd Qu.:2.000
## Max.   :2.00     Max.   :2.000    Max.   :2.000
##      i11              i12              i13
## Min.   :0.000    Min.   :1.000    Min.   :0.000
## 1st Qu.:1.000    1st Qu.:2.000    1st Qu.:2.000
## Median :2.000    Median :2.000    Median :2.000
## Mean   :1.613    Mean   :1.827    Mean   :1.693
## 3rd Qu.:2.000    3rd Qu.:2.000    3rd Qu.:2.000
## Max.   :2.000    Max.   :2.000    Max.   :2.000
##      i14              i15              i16
## Min.   :0.000    Min.   :0.00     Min.   :0.000
## 1st Qu.:1.000    1st Qu.:1.00     1st Qu.:1.000
## Median :1.000    Median :2.00     Median :1.000
## Mean   :1.173    Mean   :1.48     Mean   :1.107
## 3rd Qu.:2.000    3rd Qu.:2.00     3rd Qu.:2.000
## Max.   :2.000    Max.   :2.00     Max.   :2.000
##      i17              i18              i19
## Min.   :0.000    Min.   :0.000    Min.   :0.00
## 1st Qu.:1.000    1st Qu.:2.000    1st Qu.:2.00
## Median :1.000    Median :2.000    Median :2.00
## Mean   :1.267    Mean   :1.933    Mean   :1.88
## 3rd Qu.:2.000    3rd Qu.:2.000    3rd Qu.:2.00
## Max.   :2.000    Max.   :2.000    Max.   :2.00
##      i20              i21              i22
## Min.   :0.0000   Min.   :0.000    Min.   :0.000
## 1st Qu.:0.0000   1st Qu.:1.000    1st Qu.:1.000
## Median :1.0000   Median :2.000    Median :1.000
## Mean   :0.6667   Mean   :1.587    Mean   :1.293
## 3rd Qu.:1.0000   3rd Qu.:2.000    3rd Qu.:2.000
## Max.   :2.0000   Max.   :2.000    Max.   :2.000
##      i23              i24              i25
## Min.   :0.00     Min.   :0.000    Min.   :0.000
## 1st Qu.:0.00     1st Qu.:1.000    1st Qu.:1.000
## Median :0.00     Median :2.000    Median :1.000
## Mean   :0.56     Mean   :1.427    Mean   :1.133
```

```
## 3rd Qu.:1.00    3rd Qu.:2.000    3rd Qu.:2.000
## Max.   :2.00    Max.   :2.000    Max.   :2.000
```

From the summary of `science`, we can see that there are no missing data. We can also get a general sense of the scales, range, and distribution of each variable in the data set. We can see that Student ID numbers range from 1 to 75, and that the maximum rating on the items was $x = 2$. Importantly, we can see that for Item 12, the minimum rating was $x = 1$, and no ratings in the first category ($x = 0$) were observed in our sample. In contrast to the RSM analysis in Chapter 4, we can include this item in our analysis with the PCM because the PCM estimates rating scale category thresholds separately for each item.

Run the Partial Credit Model

Next we need to isolate the item response matrix from the descriptive variables in the data (in this case, student identification numbers). To do so, we will create an object made up of only the item responses by removing the first variable (`student`) from the data.

```
science.responses <- subset(science, select = -
    student)
```

We will use the `summary()` function to calculate descriptive statistics for the `science.responses` object to check our work and ensure that the responses are ready for analysis.

```
summary(science.responses)
```

```
##         i1                i2                i3
## Min.   :0.000    Min.   :0.000    Min.   :0.000
## 1st Qu.:1.000    1st Qu.:1.000    1st Qu.:1.000
## Median :1.000    Median :2.000    Median :1.000
## Mean   :1.453    Mean   :1.547    Mean   :1.173
## 3rd Qu.:2.000    3rd Qu.:2.000    3rd Qu.:2.000
## Max.   :2.000    Max.   :2.000    Max.   :2.000
##         i4                i5
## Min.   :0.0000   Min.   :0.0000
## 1st Qu.:0.0000   1st Qu.:0.0000
## Median :1.0000   Median :0.0000
## Mean   :0.6933   Mean   :0.4933
## 3rd Qu.:1.0000   3rd Qu.:1.0000
## Max.   :2.0000   Max.   :2.0000
##         i6                i7                i8
## Min.   :0.000    Min.   :0.00     Min.   :0.00
## 1st Qu.:1.000    1st Qu.:0.00     1st Qu.:0.00
## Median :1.000    Median :1.00     Median :1.00
## Mean   :1.213    Mean   :0.92     Mean   :0.72
```

```
## 3rd Qu.:2.000    3rd Qu.:1.50     3rd Qu.:1.00
## Max.   :2.000    Max.    :2.00    Max.    :2.00
##         i9              i10              i11
## Min.   :0.000    Min.    :0.000   Min.    :0.000
## 1st Qu.:0.000    1st Qu.:2.000    1st Qu.:1.000
## Median :1.000    Median :2.000    Median :2.000
## Mean   :1.067    Mean    :1.733   Mean    :1.613
## 3rd Qu.:2.000    3rd Qu.:2.000    3rd Qu.:2.000
## Max.   :2.000    Max.    :2.000   Max.    :2.000
##         i12             i13              i14
## Min.   :1.000    Min.    :0.000   Min.    :0.000
## 1st Qu.:2.000    1st Qu.:2.000    1st Qu.:1.000
## Median :2.000    Median :2.000    Median :1.000
## Mean   :1.827    Mean    :1.693   Mean    :1.173
## 3rd Qu.:2.000    3rd Qu.:2.000    3rd Qu.:2.000
## Max.   :2.000    Max.    :2.000   Max.    :2.000
##         i15             i16              i17
## Min.   :0.00     Min.    :0.000   Min.    :0.000
## 1st Qu.:1.00     1st Qu.:1.000    1st Qu.:1.000
## Median :2.00     Median :1.000    Median :1.000
## Mean   :1.48     Mean    :1.107   Mean    :1.267
## 3rd Qu.:2.00     3rd Qu.:2.000    3rd Qu.:2.000
## Max.   :2.00     Max.    :2.000   Max.    :2.000
##         i18             i19              i20
## Min.   :0.000    Min.    :0.00    Min.    :0.0000
## 1st Qu.:2.000    1st Qu.:2.00     1st Qu.:0.0000
## Median :2.000    Median :2.00     Median :1.0000
## Mean   :1.933    Mean    :1.88    Mean    :0.6667
## 3rd Qu.:2.000    3rd Qu.:2.00     3rd Qu.:1.0000
## Max.   :2.000    Max.    :2.00    Max.    :2.0000
##         i21             i22              i23
## Min.   :0.000    Min.    :0.000   Min.    :0.00
## 1st Qu.:1.000    1st Qu.:1.000    1st Qu.:0.00
## Median :2.000    Median :1.000    Median :0.00
## Mean   :1.587    Mean    :1.293   Mean    :0.56
## 3rd Qu.:2.000    3rd Qu.:2.000    3rd Qu.:1.00
## Max.   :2.000    Max.    :2.000   Max.    :2.00
##         i24             i25
## Min.   :0.000    Min.    :0.000
## 1st Qu.:1.000    1st Qu.:1.000
## Median :2.000    Median :1.000
## Mean   :1.427    Mean    :1.133
## 3rd Qu.:2.000    3rd Qu.:2.000
## Max.   :2.000    Max.    :2.000
```

Now we are ready to run the PCM on the Liking for Science response data. We will use the PCM() function to run the model and store the results in an object called PCM.science.

```
PCM.science <- PCM(science.responses, se = TRUE)
```

```
## Warning:
## The following items have no 0-responses:
## i12
## Responses are shifted such that lowest category
## is 0.
```

When we run the PCM on the science data, we see a message indicating that the responses for item 12 have been shifted such that the lowest category is equal to zero. We need to keep this in mind when we interpret the threshold estimates for Item 12. For this item, the lowest threshold represents the threshold between $x = 1$ and $x = 2$, whereas the lowest threshold for the other items represents the threshold between $x = 0$ and $x = 1$.

Overall Model Summary

We will request a summary of the model results using the summary() function.

```
summary(PCM.science)
```

```
##
## Results of PCM estimation:
##
## Call:  PCM(X = science.responses, se = TRUE)
##
## Conditional log-likelihood: -1165.791
## Number of iterations: 148
## Number of parameters: 48
##
## Item (Category) Difficulty Parameters (eta): with
     0.95 CI:
##           Estimate Std. Error lower CI upper CI
## i1.c2      -1.338     0.616     -2.546   -0.130
## i2.c1      -0.622     0.438     -1.480    0.237
## i2.c2      -0.763     0.432     -1.610    0.084
## i3.c1      -0.605     0.338     -1.268    0.058
## i3.c2       0.920     0.404      0.128    1.712
## i4.c1       0.683     0.270      0.155    1.212
## i4.c2       3.819     0.484      2.870    4.769
## i5.c1       1.813     0.297      1.232    2.395
## i5.c2       4.548     0.495      3.578    5.517
## i6.c1      -0.682     0.349     -1.366    0.001
```

```
## i6.c2       0.701      0.407    -0.096     1.497
## i7.c1       0.367      0.288    -0.196     0.931
## i7.c2       2.286      0.392     1.519     3.054
## i8.c1       0.723      0.273     0.188     1.257
## i8.c2       3.515      0.453     2.627     4.403
## i9.c1       0.204      0.305    -0.395     0.802
## i9.c2       1.536      0.372     0.808     2.265
## i10.c1     -2.060      0.748    -3.525    -0.594
## i10.c2     -2.610      0.727    -4.034    -1.186
## i11.c1     -2.421      0.731    -3.854    -0.989
## i11.c2     -2.252      0.725    -3.673    -0.831
## i12.c1     -0.922      0.327    -1.562    -0.281
## i13.c1     -0.163      0.517    -1.176     0.850
## i13.c2     -1.295      0.454    -2.184    -0.406
## i14.c1     -0.471      0.333    -1.123     0.182
## i14.c2      0.950      0.396     0.175     1.725
## i15.c1     -1.052      0.439    -1.913    -0.192
## i15.c2     -0.675      0.453    -1.562     0.213
## i16.c1     -0.571      0.327    -1.212     0.070
## i16.c2      1.281      0.410     0.478     2.085
## i17.c1     -0.589      0.354    -1.283     0.105
## i17.c2      0.478      0.399    -0.304     1.260
## i18.c1     -1.256      1.140    -3.491     0.979
## i18.c2     -3.879      0.993    -5.825    -1.933
## i19.c1     -2.055      1.056    -4.124     0.014
## i19.c2     -3.722      0.991    -5.665    -1.780
## i20.c1      1.014      0.275     0.475     1.554
## i20.c2      3.716      0.454     2.827     4.604
## i21.c1     -1.002      0.479    -1.941    -0.062
## i21.c2     -1.136      0.475    -2.067    -0.205
## i22.c1     -0.536      0.357    -1.237     0.165
## i22.c2      0.375      0.396    -0.401     1.151
## i23.c1      1.636      0.293     1.062     2.211
## i23.c2      4.092      0.458     3.194     4.990
## i24.c1     -0.565      0.392    -1.334     0.204
## i24.c2     -0.214      0.407    -1.011     0.584
## i25.c1     -0.154      0.318    -0.777     0.469
## i25.c2      1.196      0.381     0.449     1.943
##
## Item Easiness Parameters (beta) with 0.95 CI:
##               Estimate Std. Error lower CI upper CI
## beta i1.c1      2.246      0.602     1.067     3.426
## beta i1.c2      1.338      0.616     0.130     2.546
## beta i2.c1      0.622      0.438    -0.237     1.480
## beta i2.c2      0.763      0.432    -0.084     1.610
```

```
## beta i3.c1      0.605      0.338     -0.058      1.268
## beta i3.c2     -0.920      0.404     -1.712     -0.128
## beta i4.c1     -0.683      0.270     -1.212     -0.155
## beta i4.c2     -3.819      0.484     -4.769     -2.870
## beta i5.c1     -1.813      0.297     -2.395     -1.232
## beta i5.c2     -4.548      0.495     -5.517     -3.578
## beta i6.c1      0.682      0.349     -0.001      1.366
## beta i6.c2     -0.701      0.407     -1.497      0.096
## beta i7.c1     -0.367      0.288     -0.931      0.196
## beta i7.c2     -2.286      0.392     -3.054     -1.519
## beta i8.c1     -0.723      0.273     -1.257     -0.188
## beta i8.c2     -3.515      0.453     -4.403     -2.627
## beta i9.c1     -0.204      0.305     -0.802      0.395
## beta i9.c2     -1.536      0.372     -2.265     -0.808
## beta i10.c1     2.060      0.748      0.594      3.525
## beta i10.c2     2.610      0.727      1.186      4.034
## beta i11.c1     2.421      0.731      0.989      3.854
## beta i11.c2     2.252      0.725      0.831      3.673
## beta i12.c1     0.922      0.327      0.281      1.562
## beta i13.c1     0.163      0.517     -0.850      1.176
## beta i13.c2     1.295      0.454      0.406      2.184
## beta i14.c1     0.471      0.333     -0.182      1.123
## beta i14.c2    -0.950      0.396     -1.725     -0.175
## beta i15.c1     1.052      0.439      0.192      1.913
## beta i15.c2     0.675      0.453     -0.213      1.562
## beta i16.c1     0.571      0.327     -0.070      1.212
## beta i16.c2    -1.281      0.410     -2.085     -0.478
## beta i17.c1     0.589      0.354     -0.105      1.283
## beta i17.c2    -0.478      0.399     -1.260      0.304
## beta i18.c1     1.256      1.140     -0.979      3.491
## beta i18.c2     3.879      0.993      1.933      5.825
## beta i19.c1     2.055      1.056     -0.014      4.124
## beta i19.c2     3.722      0.991      1.780      5.665
## beta i20.c1    -1.014      0.275     -1.554     -0.475
## beta i20.c2    -3.716      0.454     -4.604     -2.827
## beta i21.c1     1.002      0.479      0.062      1.941
## beta i21.c2     1.136      0.475      0.205      2.067
## beta i22.c1     0.536      0.357     -0.165      1.237
## beta i22.c2    -0.375      0.396     -1.151      0.401
## beta i23.c1    -1.636      0.293     -2.211     -1.062
## beta i23.c2    -4.092      0.458     -4.990     -3.194
## beta i24.c1     0.565      0.392     -0.204      1.334
## beta i24.c2     0.214      0.407     -0.584      1.011
## beta i25.c1     0.154      0.318     -0.469      0.777
## beta i25.c2    -1.196      0.381     -1.943     -0.449
```

The summary of the PCM output includes the Conditional Log-likelihood statistic, details about the number of iterations and model parameters, and a table with item parameter estimates, their standard errors, and confidence intervals. It is important to note that the item parameter estimates included in this preliminary output are *item easiness* estimates—*not* item difficulty estimates. We will examine item difficulty parameter estimates in detail later in our analysis.

Wright Map

Next, we will create a Wright Map to display our model results. This plot will provide us with an overview of the distribution of person, item, and threshold parameters. We will create the plot using the *eRm* package function plotPImap() on the model object (PCM.science).

```
plotPImap(PCM.science, main = "Liking for Science
    Partial Credit Model Wright Map")
```

Liking for Science Partial Credit Model Wright Map

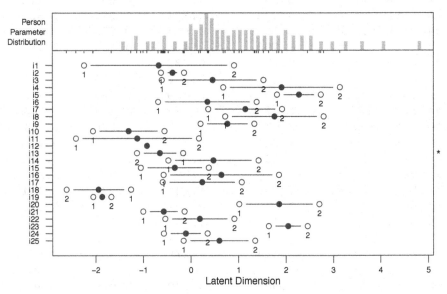

In this *Wright Map* display, the results from the PCM analysis of the Liking for Science data are summarized graphically. The figure is organized as follows:

Starting at the bottom of the figure, the horizontal axis (labeled *Latent Dimension*) is the logit scale that represents the latent variable. In the application of the Liking for Science data, lower numbers indicate less-favorable attitudes toward science activities, and higher numbers indicate more-favorable attitudes toward science activities.

The central panel of the figure shows item difficulty locations on the logit scale for the 25 Liking for Science items; the y-axis for this panel shows the item labels. By default in *eRm*, the items are ordered according to their original order in the response matrix. The items can be ordered by difficulty by adding `sorted = TRUE` as an argument in the `plotPImap()` call. For each item, a solid circle plotting symbol shows the overall location estimate. This solid circle symbol is connected to two open-circle symbols that show the locations of the rating scale category thresholds. Each threshold is labeled with either a 1 to indicate the threshold between rating scale category $x = 0$ and $x = 1$, or a 2 to indicate the threshold between rating scale category $x = 1$ and $x = 2$. Because the PCM estimates thresholds separately for each item, the distance between the thresholds varies across items. In addition, an asterisk symbol (*) is shown to the right of the central panel of the Wright map that corresponds to item 13. For this item, the thresholds were disordered.

The upper panel of the figure shows a histogram of person (in this case, children) location estimates on the logit scale. Small vertical lines on the x-axis of this histogram show the points on the logit scale at which information (variance) is maximized for the sample of persons and items in the analysis. These lines can be omitted by adding `irug = FALSE` as an argument in the `plotPImap()` function.

Item Parameters

Next, we will examine the item parameter estimates. In the PCM, item difficulty and threshold locations are combined. The *eRm* package reports item-specific rating scale category threshold parameters for each item as part of the `PCM()` function that we used to estimate the model. We extract these parameters, print them to the console, and calculate summary statistics for them with the following code.

```
item.locations <- PCM.science$etapar
item.locations
```

```
##       i1.c2        i2.c1       i2.c2        i3.c1
## -1.3377580   -0.6216355  -0.7626876   -0.6050660
##       i3.c2        i4.c1       i4.c2        i5.c1
##  0.9203065    0.6830579   3.8192173    1.8133435
##       i5.c2        i6.c1       i6.c2        i7.c1
##  4.5475811   -0.6820409   0.7007698    0.3674800
##       i7.c2        i8.c1       i8.c2        i9.c1
##  2.2864317    0.7226104   3.5152189    0.2039460
##       i9.c2       i10.c1      i10.c2       i11.c1
##  1.5361102   -2.0597760  -2.6103623   -2.4212872
##      i11.c2       i12.c1      i13.c1       i13.c2
## -2.2520112   -0.9215084  -0.1628857   -1.2952600
##      i14.c1       i14.c2      i15.c1       i15.c2
```

```
## -0.4705264   0.9503012  -1.0524982  -0.6745432
##      i16.c1       i16.c2       i17.c1       i17.c2
## -0.5711244   1.2814659  -0.5888473   0.4778861
##      i18.c1       i18.c2       i19.c1       i19.c2
## -1.2560750  -3.8790609  -2.0550593  -3.7221579
##      i20.c1       i20.c2       i21.c1       i21.c2
##  1.0143771   3.7155280  -1.0019024  -1.1358874
##      i22.c1       i22.c2       i23.c1       i23.c2
## -0.5359997   0.3749610   1.6364490   4.0922517
##      i24.c1       i24.c2       i25.c1       i25.c2
## -0.5652665  -0.2136329  -0.1540310   1.1958751
```

```
summary(item.locations)
```

```
##     Min. 1st Qu.   Median     Mean 3rd Qu.     Max.
## -3.8791  -1.0146  -0.3421   0.0468   0.9663   4.5476
```

Because of the nature of the estimation process used in *eRm*, the item.locations object that we just created does not include the location estimate for the first threshold for the first item. One can calculate the location for item 1, threshold 1, by subtracting the sum of the item parameter estimates from zero. In the following code, we find the location for item 1 threshold 1, and then create a new object with all of the item + threshold locations.

```
i1 <- 0 - sum(item.locations[(1:length(item.locations
   )) - 1])
item.locations.all <- c(i1, item.locations[(1:length(
   item.locations)) - 1])
item.locations.all[1]
```

```
##
## -1.050402
```

Alternatively, one can apply the `thresholds()` function to the model object to find the item locations from the PCM. This procedure provides item location estimates (δ) as well as estimates of the item location combined with rating scale category thresholds ($\delta + \tau$). However, the values produced with this function are not centered at zero logits, so a little manipulation is required to obtain the centered values. In the following code, we apply the `thresholds()` function to obtain the uncentered item location estimates and then calculate centered item locations.

```
# Apply thresholds() function to the model object in
   order to obtain item locations (not centered at
   zero logits):
items.and.taus <- thresholds(PCM.science)
```

```
items.and.taus.table <- as.data.frame(items.and.taus$
   threshtable)
uncentered.item.locations <- items.and.taus.table$X1.
   Location

# Set the mean of the item locations to zero logits:
centered.item.locations <- scale(uncentered.item.
   locations, scale = FALSE)

# Summarize the results:
summary(centered.item.locations)

##          V1
##  Min.   :-2.13328
##  1st Qu.:-0.84138
##  Median : 0.04519
##  Mean   : 0.00000
##  3rd Qu.: 0.57430
##  Max.   : 2.08004
```

It is important to note that the eta parameter estimates from the PCM object include *cumulative* rating scale category thresholds. Cumulative thresholds are not used in all Rasch model applications. Instead, many researchers use Rasch-Andrich (i.e., adjacent categories) thresholds (Andrich, 2015; Mellenbergh, 1995). We will calculate the values of the adjacent-categories rating scale category thresholds using the results from the `thresholds()` function.

In our example, we have a rating scale with three categories, so we have two rating scale category thresholds for all of the items in which all categories were observed. In the following code chunk, we create an empty object in which to store the threshold estimates (`tau.matrix`), and then we use two for-loops to calculate the estimates and store them in our object.

```
# Specify the number of items that were included in
   the analysis:
n.items <- ncol(science.responses)

# Specify the number of thresholds as the maximum
   observed score in the response matrix (be sure the
   responses begin at category 0):
n.thresholds <- max(science.responses)

# Create a matrix in which to store the adjacent-
   category threshold values for each item:
tau.matrix <- matrix(data = NA, ncol = n.thresholds,
   nrow = n.items)
```

```
# Calculate adjacent-category threshold values:
for(item.number in 1:n.items){
  for(tau in 1:n.thresholds){
    tau.matrix[item.number, tau] <- (items.and.taus.
      table[item.number, (1+tau)] -
                                    items.and.taus
                                      .table[item
                                      .number,1])
                                      [1]

  }
}
```

We can examine the threshold results for evidence that the rating scale categories are ordered as expected according to the ordinal rating scale. When scale categories are ordered as expected, the first threshold has a lower location on the latent variable compared to the second threshold, and so on. In our example with the Liking for Science data, we have two thresholds, so we need to evaluate the expression $\tau_1 <= \tau_2$. We check this property using the following code, which compares the first threshold estimate (saved in the first column of tau.matrix) to the second threshold estimate (saved in the second column of tau.matrix). Analysts whose scales include more rating scale categories should evaluate this property for all pairs of adjacent thresholds.

```
tau.matrix[,1] <= tau.matrix[,2]
```

```
##  [1]  TRUE  TRUE  TRUE  TRUE  TRUE  TRUE  TRUE
##  [8]  TRUE  TRUE  TRUE  TRUE    NA FALSE  TRUE
## [15]  TRUE  TRUE  TRUE FALSE  TRUE  TRUE  TRUE
## [22]  TRUE  TRUE  TRUE  TRUE
```

For Item 12, the expression returns a value of "NA" because only one threshold was estimated for this item. The results indicate that $\tau_1 <= \tau_2$ is true for all items except Item 13. The disordered thresholds for Item 13 indicate that lower locations on the construct were required to provide a rating in category 1 compared to category 2, when the opposite order would have been expected given the rating scale category labels. This result suggests that additional research may be required to understand students' responses to Item 13.

Next, we will calculate standard errors for each item + threshold location and store them in an object called delta.tau.se. We will use summary() to examine descriptive statistics for each item + threshold standard error.

```
delta.tau.se <- items.and.taus$se.thresh
summary(delta.tau.se)
```

```
##      Min. 1st Qu.  Median    Mean 3rd Qu.    Max.
##    0.2697  0.2968  0.3270  0.4005  0.4380  1.1405
```

Item Response Functions

We will examine graphical displays of item difficulty using item response functions. With the PCM, the *eRm* package creates plots of the probability for a rating in each rating scale category, conditional on person locations on the latent variable. In the following code, we use `plotICC()` from *eRm* to create rating scale category probability plots for the first 5 items in our analysis. We included `ask = FALSE` in our function call to generate all of the plots at once. Readers who need to specify plots for different items can update the `item.subset =` specification to produce the desired plots.

```
plotICC(PCM.science, ask = FALSE, item.subset = c
    (1:5))
```

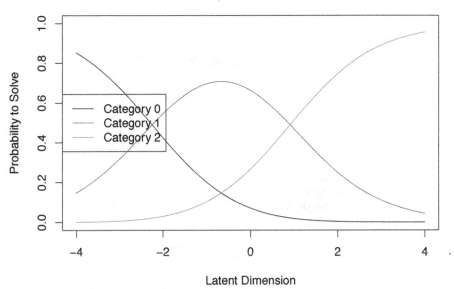

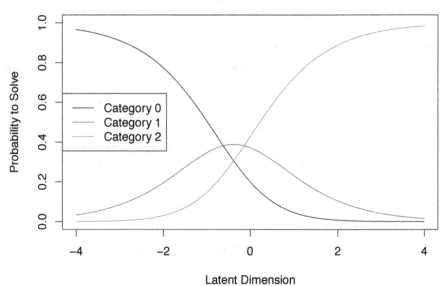

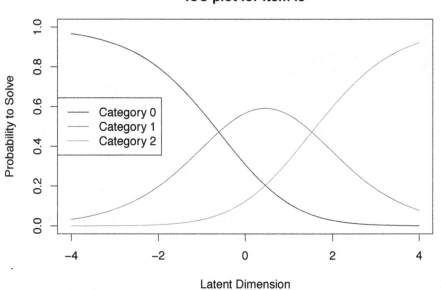

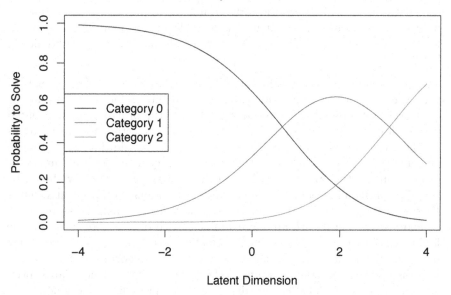

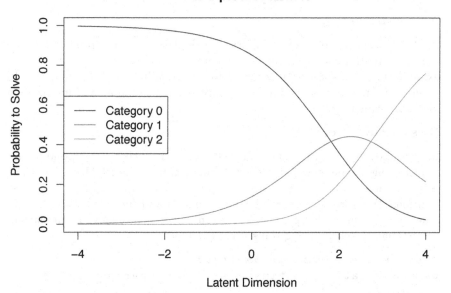

In each item-specific plot, the x-axis is the logit scale that represents the latent variable; in our example, this scale represents favorability toward science activities. The y-axis is the probability for a rating in each category, conditional on person locations on the latent variable. Separate lines in different colors show the conditional probability for each category of responses observed for the item of interest. Because we used the PCM, the overall shape of the curves and the relative distance between the curves is unique for each items—reflecting the individual set of rating scale category thresholds for each item. In addition, the location of the curves on the x-axis shifts to reflect each item's overall difficulty level.

To supplement the numeric evaluation of rating scale category ordering, analysts can examine plots of rating scale category probabilities for evidence that the categories are ordered as expected. Specifically, the probabilities associated with each item should increase and decrease in the expected order as locations on the latent variable progress from low to high. In addition, analysts can examine these plots for evidence that each category describes a unique range of locations on the latent variable. In other words, each category should be the most-probable response for some range of locations on the latent variable. Distinct category probability curves provide evidence that the rating scale categories can distinguish among examinees with different locations on the latent variable. In cases where such distinctions are not observed, researchers may consider revising the scale to include fewer categories prior to further analysis, as well as making revisions to the scale length or category labels prior to future administrations. For details about these procedures and guidance in decision-making related to rating scale revisions, we recommend that readers consult Bond et al. (2019), chapter 11, as well as Linacre (2002).

Note about item fit

In the *eRm* package, it is necessary to calculate person parameters *before* item fit statistics can be calculated. Accordingly, we will proceed with a brief examination of person parameters before we conduct item fit analyses. In practice, we recommend examining item fit before examining and interpreting location estimates in detail.

Person Parameters

In the following code, we calculate person locations that correspond to our model using the `person.parameter()` function with the PCM model object (`PCM.science`). This function also produces standard errors for the person locations. We store the person location estimates and their standard errors in a new data frame called `person.locations`, and then request a summary of the estimation results using the `summary()` function.

```
# Calculate person parameters:
person.locations.estimate <- person.parameter(PCM.
    science)

# Store person parameters and their standard errors
    in a dataframe object:
person.locations <- cbind.data.frame(person.locations
    .estimate$thetapar,
```

```
                                          person.locations
                                            .estimate$se.
                                                   theta)
names(person.locations) <- c("theta", "SE")
```

```
# View summary statistics for person parameters:
summary(person.locations)
```

```
##        theta                    SE
##   Min.    :-1.4184    Min.     :0.3312
##   1st Qu.: 0.2206    1st Qu.:0.3346
##   Median : 0.7311    Median :0.3470
##   Mean    : 0.9645    Mean     :0.3802
##   3rd Qu.: 1.6574    3rd Qu.:0.3787
##   Max.    : 4.8114    Max.     :1.0280
```

The estimation procedure in *eRm* does not directly produce parameter estimates for persons with extreme scores. In our example, extreme scores would result from a child giving a response of $x = 0$ to all items or a child giving a response of $x = 2$ to all items. For these children, a standard error is not calculated. In our example, Child 2 had an extreme score because they gave a rating of 2 to all items.

Item Fit

Next, we will conduct a brief exploration of item fit statistics for the Liking for Science items. We considered item fit in detail in Chapter 3; readers can use the procedures in that chapter to examine item fit in detail for the PCM.

To calculate numeric item fit statistics, we will use the function itemfit() from *eRm* on the person parameter object (**person.locations.estimate**). This function produces several item fit statistics, including infit mean square error (MSE), outfit MSE, and standardized infit and outfit statistics. We will store the item fit results in a new object called **item.fit**, and then format this object as a dataframe for easy manipulation and exporting.

```
item.fit.results <- itemfit(person.locations.estimate
    )
```

```
item.fit <- cbind.data.frame(item.fit.results$i.
    infitMSQ ,
                              item.fit.results$i.
                                 outfitMSQ ,
                              item.fit.results$i.
                                 infitZ ,
                              item.fit.results$i.
                                 outfitZ)
```

```
names(item.fit) <- c("infit_MSE", "outfit_MSE", "std_
    infit", "std_outfit")
```

Next, we will request a summary of the numeric item fit statistics using the *summary()* function.

```
summary(item.fit)
```

```
##      infit_MSE            outfit_MSE
##   Min.    :0.6377     Min.    :0.5321
##   1st Qu.:0.7291      1st Qu.:0.5917
##   Median :0.8068      Median :0.7034
##   Mean    :0.9696     Mean    :1.0642
##   3rd Qu.:1.0584      3rd Qu.:1.0185
##   Max.    :2.2781     Max.    :4.3556
##      std_infit           std_outfit
##   Min.    :-2.8262     Min.    :-2.6664
##   1st Qu.:-1.7769      1st Qu.:-1.6258
##   Median :-1.0462      Median :-1.1723
##   Mean    :-0.4437     Mean    :-0.1371
##   3rd Qu.: 0.3955      3rd Qu.: 0.2610
##   Max.    : 5.5773     Max.    : 8.1725
```

The item.fit object includes mean square error (MSE) and standardized (Z) versions of the infit and outfit statistics for each item included in the analysis. These statistics are summaries of the residuals associated with each item. When data fit Rasch model expectations, the MSE versions of infit and outfit are expected to be close to 1.00 and the standardized versions of infit and outfit are expected to be around 0.00. Please refer to Chapter 3 for a detailed discussion of item fit.

Person Fit

Next, we will conduct a brief exploration of person fit statistics. To calculate numeric person fit statistics, we will apply the function personfit() from *eRm* to the person parameter object (person.locations.estimate). This function produces several person fit statistics, including infit MSE, outfit MSE, and standardized infit and outfit MSE statistics. We will store the person fit results in a new object called person.fit, and then format this object as a dataframe for easy manipulation and exporting.

```
person.fit.results <- personfit(person.locations.
    estimate)
```

```
person.fit <- cbind.data.frame(person.fit.results$p.
    infitMSQ,
```

```
                            person.fit.results$p.
                               outfitMSQ,
                            person.fit.results$p.
                               infitZ,
                            person.fit.results$p.
                               outfitZ)

names(person.fit) <- c("infit_MSE", "outfit_MSE", "
   std_infit", "std_outfit")
```

Next, we will request a summary of the numeric person fit statistics using the summary() function.

```
summary(person.fit)
```

```
##      infit_MSE             outfit_MSE
##   Min.    :0.2637    Min.    :0.2826
##   1st Qu.:0.6797    1st Qu.:0.5444
##   Median :0.8655    Median :0.7222
##   Mean    :0.9693    Mean    :1.0642
##   3rd Qu.:1.0969    3rd Qu.:1.1500
##   Max.    :3.1714    Max.    :5.0035
##      std_infit             std_outfit
##   Min.    :-3.9448    Min.    :-2.8147
##   1st Qu.:-1.1692    1st Qu.:-0.8651
##   Median :-0.3302    Median :-0.3226
##   Mean    :-0.2348    Mean    : 0.1455
##   3rd Qu.: 0.4059    3rd Qu.: 0.5419
##   Max.    : 5.1201    Max.    : 5.6258
```

The **person.fit** object includes MSE and standardized (Z) versions of the infit and outfit statistics for each person.

Next, we calculate the reliability of separation statistics for persons and items.

```
## Person separation reliability
person.separation.reliability <- SepRel(person.
   locations.estimate)
person.separation.reliability
```

```
##
## Separation Reliability: 0.8912

## Item separation reliability:

# Get Item scores
ItemScores <- colSums(science.responses)
```

```
# Get Item SD
ItemSD <- apply(science.responses,2,sd)

# Calculate the SE of the Item
ItemSE <- ItemSD/sqrt(length(ItemSD))

# Compute the Observed Variance (also known as Total
    Person Variability or Squared Standard Deviation)
SSD.ItemScores <- var(ItemScores)

# Compute the Mean Square Measurement error (also
    known as Model Error variance)
Item.MSE <- sum((ItemSE)^2) / length(ItemSE)

# Compute the Item Separation Reliability
item.separation.reliability <- (SSD.ItemScores-Item.
    MSE) / SSD.ItemScores

item.separation.reliability

## [1] 0.9999826
```

5.3   Summarize the Results in Tables

As a final step, we will create tables that summarize the calibrations of the persons, items, and rating scale category thresholds.

Table 1 provides an overview of the logit-scale locations, standard errors, fit statistics, and reliability statistics for items and persons. This type of table is useful for reporting the results from Rasch model analyses because it provides a succinct overview of the location estimates and numeric model-data fit statistics for the items and persons in the analysis.

```
PCM_summary.table.statistics <- c("Logit Scale
    Location Mean",
                                  "Logit Scale Location
                                      SD",
                                  "Standard Error Mean",
                                  "Standard Error SD",
                                  "Outfit MSE Mean",
                                  "Outfit MSE SD",
```

```
                              "Infit MSE Mean",
                              "Infit MSE SD",
                              "Std. Outfit Mean",
                              "Std. Outfit SD",
                              "Std. Infit Mean",
                              "Std. Infit SD",
                              "Separation.reliability
                                 ")

PCM_item.summary.results <- rbind(mean(centered.item.
    locations),
                              sd(centered.item.
                                 locations),
                              mean(delta.tau.se),
                              sd(delta.tau.se),
                              mean(item.fit.results$i
                                 .outfitMSQ),
                              sd(item.fit.results$i.
                                 outfitMSQ),
                              mean(item.fit.results$i
                                 .infitMSQ),
                              sd(item.fit.results$i.
                                 infitMSQ),
                              mean(item.fit.results$i
                                 .outfitZ),
                              sd(item.fit.results$i.
                                 outfitZ),
                              mean(item.fit.results$i
                                 .infitZ),
                              sd(item.fit.results$i.
                                 infitZ),
                              item.separation.
                                 reliability)

PCM_person.summary.results <- rbind(mean(person.
    locations$theta),
                              sd(person.locations$
                                 theta),
                              mean(person.locations$
                                 SE),
                              sd(person.locations$SE)
                                 ,
                              mean(person.fit$outfit_
                                 MSE),
```

```
                              sd(person.fit$outfit_
                                 MSE),
                              mean(person.fit$infit_
                                 MSE),
                              sd(person.fit$infit_MSE
                                 ),
                              mean(person.fit$std_
                                 outfit),
                              sd(person.fit$std_
                                 outfit),
                              mean(person.fit$std_
                                 infit),
                              sd(person.fit$std_infit
                                 ),
                              as.numeric(person.
                                 separation.
                                 reliability))

# Round the values for presentation in a table:
PCM_item.summary.results_rounded <- round(PCM_item.
   summary.results, digits = 2)

PCM_person.summary.results_rounded <- round(PCM_
   person.summary.results, digits = 2)

PCM_Table1 <- cbind.data.frame(PCM_summary.table.
   statistics,
                     PCM_item.summary.results_
                        rounded,
                     PCM_person.summary.results
                        _rounded[,1])

# Add descriptive column labels:
names(PCM_Table1) <- c("Statistic", "Items", "Persons
   ")
```

Table 2 summarizes the overall calibrations of individual items. For data sets
with manageable sample sizes such as the Liking for Science data example
in this chapter, we recommend reporting details about each item in a table
similar to this one.

```
# Calculate the average rating for each item:
Avg_Rating <- apply(science.responses, 2, mean)
```

```
# Combine item calibration results in a table:

PCM_Table2 <- cbind.data.frame(c(1:ncol(science.
    responses)),
                                Avg_Rating,
                                centered.item.locations,
                                tau.matrix,
                                item.fit$outfit_MSE,
                                item.fit$std_outfit,
                                item.fit$infit_MSE,
                                item.fit$std_infit)

# Add descriptive column labels:
names(PCM_Table2) <- c("Task ID", "Average Rating", "
    Item Location","Threshold 1", "Threshold 2", "
    Outfit MSE","Std. Outfit", "Infit MSE","Std. Infit
    ")

# Sort Table 2 by Item difficulty:
PCM_Table2 <- PCM_Table2[order(-PCM_Table2$`Item
    Location`),]

# Round the numeric values (all columns except the
    first one) to 2 digits:
PCM_Table2[, -1] <- round(PCM_Table2[,-1], digits =
    2)
```

Finally, Table 3 summarizes person calibrations. When there is a relatively large person sample size, it may be more useful to present the results as they relate to individual persons or subsets of the person sample as they are relevant to the purpose of the analysis.

In our person calibration table, we have included the results for all of the children with non-extreme scores. This includes all of the children in our sample except Child #2.

```
# Calculate average rating for persons who did not
    have extreme scores
Person_Avg_Rating <- apply(person.locations.estimate$
    X.ex,1, mean)

# Combine person calibration results in a table:
PCM_Table3 <- cbind.data.frame(rownames(person.
    locations),
                                Person_Avg_Rating,
                                person.locations$theta,
```

```
                    person.locations$SE,
                    person.fit$outfit_MSE,
                    person.fit$std_outfit,
                    person.fit$infit_MSE,
                    person.fit$std_infit)

# Add descriptive column labels:
names(PCM_Table3) <- c("Child ID", "Average Rating",
    "Person Location","Person SE","Outfit MSE","Std.
    Outfit", "Infit MSE","Std. Infit")

# Round the numeric values (all columns except the
    first one) to 2 digits:
PCM_Table3[, -1] <- round(PCM_Table3[,-1], digits =
    2)
```

5.4   PCM Application with MMLE in *TAM*

The next section includes an illustration of PCM analyses with the *Test Analysis Modules* or *TAM* package (Robitzsch et al., 2021) with MMLE. After this illustration, we also demonstrate the use of *TAM* to estimate the PCM with JMLE. These illustrations use the Liking for Science data set that we described earlier.

Except where there are significant differences between the *eRm* and *TAM* procedures, we provide fewer details about the analysis procedures and interpretations in this section compared to the first illustration.

Prepare for the Analyses

To get started with the *TAM* package, install and load it into your R environment using the following code.

```
#install.packages("TAM")
library("TAM")
```

To facilitate the example analysis, we will also use the *WrightMap* package (Irribarra and Freund, 2014):

```
# install.packages("WrightMap")
library("WrightMap")
```

If you have not already imported the Liking for Science data and isolated the response matrix, please do so before continuing with the *TAM* analyses.

Run the Partial Credit Model

To obtain Rasch-Andrich thresholds (i.e., adjacent-categories thresholds) from our analysis, we need to generate a design matrix for the model that includes specifications for those parameters. We will do this using the designMatrices () function from *TAM* and save the result in a new object called design. matrix.

```
design.matrix <- designMatrices(resp=science.
    responses, modeltype="PCM", constraint = "items")$
    A
```

Now we can run the PCM with our design matrix using the tam.mml() function with several specifications.

```
PCM.science_MMLE <- tam.mml(science.responses,
    irtmodel="PCM", A = design.matrix, constraint = "
    items", verbose = FALSE)
```

Next, we will request a summary of the model results using the summary() function.

```
summary(PCM.science_MMLE)
```

```
## -----------------------------------------------------
## TAM 3.7-16 (2021-06-24 14:31:37)
## R version 4.1.0 (2021-05-18) x86_64, darwin17.0 |
    nodename=XXXX | login=root
##
## Date of Analysis: 2021-10-27 13:46:46
## Time difference of 0.03634787 secs
## Computation time: 0.03634787
##
## Multidimensional Item Response Model in TAM
##
## IRT Model: PCM2
## Call:
## tam.mml(resp = science.responses, irtmodel = "PCM
    ", constraint = "items",
##      A = design.matrix, verbose = FALSE)
##
## -----------------------------------------------------
## Number of iterations = 51
## Numeric integration with 21 integration points
##
## Deviance = 2870.59
## Log likelihood = -1435.29
```

```
## Number of persons = 75
## Number of persons used = 75
## Number of items = 25
## Number of estimated parameters = 52
##      Item threshold parameters = 50
##      Item slope parameters = 0
##      Regression parameters = 1
##      Variance/covariance parameters = 1
##
## AIC = 2975  | penalty=104     | AIC=-2*LL + 2*p
## AIC3 = 3027 | penalty=156     | AIC3=-2*LL + 3*p
## BIC = 3095  | penalty=224.51    | BIC=-2*LL + log(
   n)*p
## aBIC = 2928 | penalty=57.85     | aBIC=-2*LL + log
   ((n-2)/24)*p  (adjusted BIC)
## CAIC = 3147 | penalty=276.51    | CAIC=-2*LL + [
   log(n)+1]*p  (consistent AIC)
## AICc = 3225 | penalty=354.55    | AICc=-2*LL + 2*
   p + 2*p*(p+1)/(n-p-1)  (bias corrected AIC)
## GHP = 0.79322    | GHP=( -LL + p ) / (#Persons *
   #Items)  (Gilula-Haberman log penalty)
##
## -----------------------------------------------
## EAP Reliability
## [1] 0.9
## -----------------------------------------------
## Covariances and Variances
##        [,1]
## [1,] 1.278
## -----------------------------------------------
## Correlations and Standard Deviations (in the
   diagonal)
##        [,1]
## [1,] 1.13
## -----------------------------------------------
## Regression Coefficients
##          [,1]
## [1,] 0.05447
## -----------------------------------------------
## Item Parameters -A*Xsi
##    item  N     M xsi.item AXsi_.Cat1 AXsi_.Cat2
## 1    i1 75 1.453   -1.609     -3.211     -3.217
## 2    i2 75 1.547   -1.319     -1.584     -2.638
## 3    i3 75 1.173   -0.464     -1.539     -0.929
## 4    i4 75 0.693    0.954     -0.221      1.908
```

##	5	i5	75	0.493	1.319	0.919	2.637
##	6	i6	75	1.213	-0.575	-1.619	-1.149
##	7	i7	75	0.920	0.223	-0.544	0.446
##	8	i8	75	0.720	0.815	-0.181	1.631
##	9	i9	75	1.067	-0.148	-0.716	-0.295
##	10	i10	75	1.733	-2.263	-3.051	-4.526
##	11	i11	75	1.613	-2.075	-3.401	-4.150
##	12	i12	75	1.827	-18.706	-35.556	-37.411
##	13	i13	75	1.693	-1.594	-1.133	-3.189
##	14	i14	75	1.173	-0.448	-1.403	-0.896
##	15	i15	75	1.480	-1.272	-2.011	-2.544
##	16	i16	75	1.107	-0.286	-1.501	-0.571
##	17	i17	75	1.267	-0.685	-1.528	-1.371
##	18	i18	75	1.933	-2.931	-2.298	-5.863
##	19	i19	75	1.880	-2.841	-3.079	-5.682
##	20	i20	75	0.667	0.917	0.115	1.834
##	21	i21	75	1.587	-1.509	-1.968	-3.018
##	22	i22	75	1.293	-0.737	-1.477	-1.474
##	23	i23	75	0.560	1.106	0.742	2.212
##	24	i24	75	1.427	-1.037	-1.517	-2.074
##	25	i25	75	1.133	-0.322	-1.082	-0.643

##		B.Cat1.Dim1	B.Cat2.Dim1
##	1	1	2
##	2	1	2
##	3	1	2
##	4	1	2
##	5	1	2
##	6	1	2
##	7	1	2
##	8	1	2
##	9	1	2
##	10	1	2
##	11	1	2
##	12	1	2
##	13	1	2
##	14	1	2
##	15	1	2
##	16	1	2
##	17	1	2
##	18	1	2
##	19	1	2
##	20	1	2
##	21	1	2
##	22	1	2
##	23	1	2

```
## 24                    1              2
## 25                    1              2
##
## Item Parameters Xsi
##                    xsi         se.xsi
## i1_Cat1        -3.211          0.612
## i1_Cat2        -0.006          0.260
## i2_Cat1        -1.584          0.414
## i2_Cat2        -1.055          0.276
## i3_Cat1        -1.539          0.335
## i3_Cat2         0.610          0.280
## i4_Cat1        -0.221          0.264
## i4_Cat2         2.129          0.406
## i5_Cat1         0.919          0.279
## i5_Cat2         1.719          0.423
## i6_Cat1        -1.619          0.345
## i6_Cat2         0.470          0.275
## i7_Cat1        -0.544          0.278
## i7_Cat2         0.990          0.312
## i8_Cat1        -0.181          0.265
## i8_Cat2         1.812          0.378
## i9_Cat1        -0.716          0.292
## i9_Cat2         0.421          0.286
## i10_Cat1       -3.051          0.748
## i10_Cat2       -1.475          0.296
## i11_Cat1       -3.401          0.741
## i11_Cat2       -0.749          0.267
## i12_Cat1      -35.556    5931641.602
## i12_Cat2       -1.855          0.326
## i13_Cat1       -1.133          0.447
## i13_Cat2       -2.055          0.319
## i14_Cat1       -1.403          0.327
## i14_Cat2        0.507          0.278
## i15_Cat1       -2.011          0.431
## i15_Cat2       -0.534          0.265
## i16_Cat1       -1.501          0.325
## i16_Cat2        0.929          0.292
## i17_Cat1       -1.528          0.347
## i17_Cat2        0.158          0.269
## i18_Cat1       -2.298          1.049
## i18_Cat2       -3.565          0.538
## i19_Cat1       -3.079          1.041
## i19_Cat2       -2.603          0.397
## i20_Cat1        0.115          0.265
## i20_Cat2        1.719          0.381
```

```
## i21_Cat1   -1.968        0.463
## i21_Cat2   -1.050        0.276
## i22_Cat1   -1.477        0.348
## i22_Cat2    0.003        0.268
## i23_Cat1    0.742        0.275
## i23_Cat2    1.470        0.391
## i24_Cat1   -1.517        0.377
## i24_Cat2   -0.557        0.267
## i25_Cat1   -1.082        0.309
## i25_Cat2    0.439        0.281
##
## Item Parameters in IRT parameterization
##      item alpha      beta tau.Cat1 tau.Cat2
## 1     i1    1   -1.609   -1.603    1.603
## 2     i2    1   -1.319   -0.264    0.264
## 3     i3    1   -0.464   -1.074    1.074
## 4     i4    1    0.954   -1.175    1.175
## 5     i5    1    1.319   -0.400    0.400
## 6     i6    1   -0.575   -1.044    1.044
## 7     i7    1    0.223   -0.767    0.767
## 8     i8    1    0.815   -0.996    0.996
## 9     i9    1   -0.148   -0.569    0.569
## 10   i10    1   -2.263   -0.788    0.788
## 11   i11    1   -2.075   -1.326    1.326
## 12   i12    1  -18.706  -16.851   16.851
## 13   i13    1   -1.594    0.461   -0.461
## 14   i14    1   -0.448   -0.955    0.955
## 15   i15    1   -1.272   -0.739    0.739
## 16   i16    1   -0.286   -1.215    1.215
## 17   i17    1   -0.685   -0.843    0.843
## 18   i18    1   -2.931    0.633   -0.633
## 19   i19    1   -2.841   -0.238    0.238
## 20   i20    1    0.917   -0.802    0.802
## 21   i21    1   -1.509   -0.459    0.459
## 22   i22    1   -0.737   -0.740    0.740
## 23   i23    1    1.106   -0.364    0.364
## 24   i24    1   -1.037   -0.480    0.480
## 25   i25    1   -0.322   -0.760    0.760
```

The summary of the PCM output includes details about the number of iterations, global model fit statistics, a summary of the model parameters, and several other statistics.

Item Parameters

Next, we will examine the item difficulty location and rating scale category threshold estimates. As of this writing, the *TAM* package does not provide centered item estimates (mean set to zero logits) with the design matrix that we specified. As a result, we need to manually center the item locations at zero logits for ease in interpretation. We will use the same procedure that we used earlier to do the centering.

First, we need to extract the item location estimates from the model object (PCM .science_MMLE). The item locations are stored in the item_irt component of the model object, so we will extract them using the $ operator. The item_irt table includes other item parameters as well, including threshold estimates. We will examine those later. For now, we will save the overall item locations in an object called items_MMLE.

```
items_MMLE <- PCM.science_MMLE$item_irt$beta
```

Next, we will center the item parameter estimates that are stored in items_MMLE at zero logits, and request summary statistics for the estimates to check our work.

```
uncentered.item.locations_MMLE <- items_MMLE

centered.item.locations_MMLE <- scale(uncentered.item
    .locations_MMLE, scale = FALSE)

summary(centered.item.locations_MMLE)
```

```
##           V1
##   Min.   :-17.2861
##   1st Qu.: -0.1748
##   Median :  0.7340
##   Mean   :  0.0000
##   3rd Qu.:  1.2718
##   Max.   :  2.7382
```

We need to find the rating scale category threshold parameter estimates for our rating scale. In our example with the Liking for Science data, the rating scale has three categories, so there are two threshold parameters for each item in which all three categories are observed. The following code extracts the adjacent-categories threshold parameters from the item_ irt component of the PCM.science_MMLE object, and then stores them in an object called tau.estimates_MMLE.

```
# View the item parameter table (note that the
    overall item location shown here is not centered):
PCM.science_MMLE$item_irt
```

```
##      item alpha        beta    tau.Cat1    tau.Cat2
## 1    i1     1  -1.6085091  -1.6026204   1.6026204
## 2    i2     1  -1.3190982  -0.2644808   0.2644808
## 3    i3     1  -0.4644297  -1.0742282   1.0742282
## 4    i4     1   0.9539572  -1.1749690   1.1749690
## 5    i5     1   1.3187275  -0.3998253   0.3998253
## 6    i6     1  -0.5745648  -1.0440676   1.0440676
## 7    i7     1   0.2229871  -0.7672589   0.7672589
## 8    i8     1   0.8153986  -0.9962125   0.9962125
## 9    i9     1  -0.1475995  -0.5688750   0.5688750
## 10   i10    1  -2.2628393  -0.7877179   0.7877179
## 11   i11    1  -2.0751341  -1.3262764   1.3262764
## 12   i12    1 -18.7055345 -16.8507035  16.8507035
## 13   i13    1  -1.5942784   0.4610256  -0.4610256
## 14   i14    1  -0.4478096  -0.9552837   0.9552837
## 15   i15    1  -1.2721835  -0.7386319   0.7386319
## 16   i16    1  -0.2857395  -1.2150325   1.2150325
## 17   i17    1  -0.6854428  -0.8430160   0.8430160
## 18   i18    1  -2.9313161   0.6334596  -0.6334596
## 19   i19    1  -2.8409412  -0.2378013   0.2378013
## 20   i20    1   0.9168055  -0.8022383   0.8022383
## 21   i21    1  -1.5091272  -0.4592362   0.4592362
## 22   i22    1  -0.7370717  -0.7402170   0.7402170
## 23   i23    1   1.1061750  -0.3642532   0.3642532
## 24   i24    1  -1.0371253  -0.4803080   0.4803080
## 25   i25    1  -0.3215084  -0.7600188   0.7600188

# Save only the threshold estimates, which begin in
   column 4:
tau.estimates_MMLE <- PCM.science_MMLE$item_irt[, c(4
   : (3+n.thresholds))]

# View the threshold estimates:
tau.estimates_MMLE

##        tau.Cat1    tau.Cat2
## 1    -1.6026204   1.6026204
## 2    -0.2644808   0.2644808
## 3    -1.0742282   1.0742282
## 4    -1.1749690   1.1749690
## 5    -0.3998253   0.3998253
## 6    -1.0440676   1.0440676
## 7    -0.7672589   0.7672589
## 8    -0.9962125   0.9962125
## 9    -0.5688750   0.5688750
```

```
## 10    -0.7877179   0.7877179
## 11    -1.3262764   1.3262764
## 12   -16.8507035  16.8507035
## 13     0.4610256  -0.4610256
## 14    -0.9552837   0.9552837
## 15    -0.7386319   0.7386319
## 16    -1.2150325   1.2150325
## 17    -0.8430160   0.8430160
## 18     0.6334596  -0.6334596
## 19    -0.2378013   0.2378013
## 20    -0.8022383   0.8022383
## 21    -0.4592362   0.4592362
## 22    -0.7402170   0.7402170
## 23    -0.3642532   0.3642532
## 24    -0.4803080   0.4803080
## 25    -0.7600188   0.7600188
```

Note that the threshold values shown in the `tau.estimates_MMLE` must be added to the overall item location to obtain the location on the logit scale at which there is an equal probability for a rating in the corresponding adjacent categories. These values correspond to the graphical displays of rating scale category probabilities for the PCM.

As we discussed in the *eRm* illustration earlier in this chapter, it is important to examine threshold parameters for each item to ensure that the rating scale categories are ordered as expected, and that there is sufficient distance between the thresholds to describe a distinct range of locations on the latent variable.

Item Response Functions

Next, we will examine rating scale category probability plots for the items in our analysis. To save space, we have only printed the plots for the first five items here. Analysts can request plots for different items using the `items =` argument in the `plot()` function.

```
plot(PCM.science_MMLE, type="items", items = c(1:5))
```

```
## Iteration in WLE/MLE estimation  1    | Maximal
      change   1.1208
## Iteration in WLE/MLE estimation  2    | Maximal
      change   0.1558
## Iteration in WLE/MLE estimation  3    | Maximal
      change   0.0048
## Iteration in WLE/MLE estimation  4    | Maximal
      change   2e-04
## Iteration in WLE/MLE estimation  5    | Maximal
      change   0
```

```
## ----
##   WLE Reliability= 0.89
```

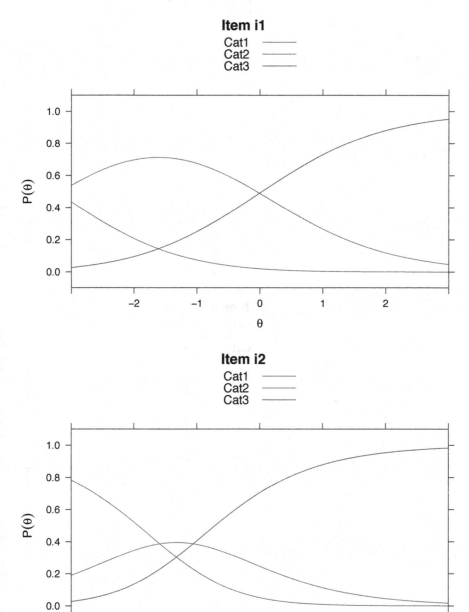

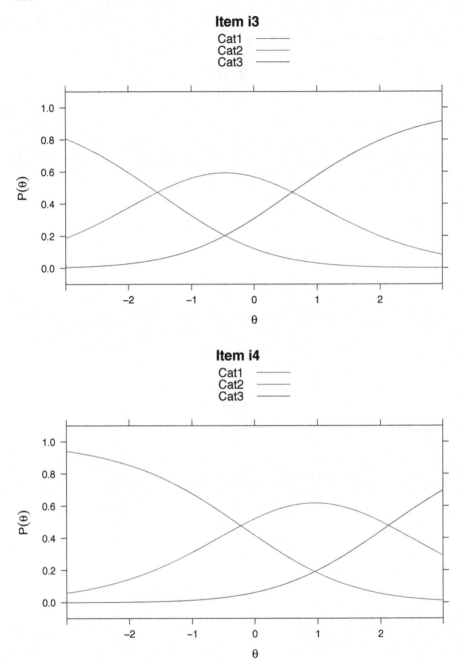

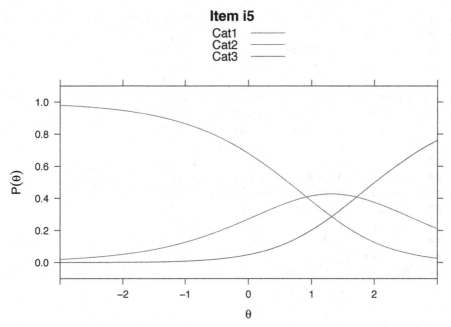

This code generates plots of rating scale category probabilities for each item. These plots have the same interpretation as the rating scale category probability plots that we generated using *eRm*, where the x-axis is the logit scale that represents the latent variable, the y-axis is the probability for a rating in each category, and individual lines show the conditional probability for a rating in each category. As we discussed earlier, analysts can examine these plots for evidence of category ordering and distinctiveness.

Item Fit

Next, we will examine numeric and graphical item fit indices using the `itemfit` () function from *TAM*. We will save the results in an object called `item.fit_MMLE`.

```
MMLE_fit <- msq.itemfit(PCM.science_MMLE)
item.fit_MMLE <- MMLE_fit$itemfit
```

Next, we will view summary statistics for the fit statistics.

```
summary(item.fit_MMLE)
```

```
##       item              fitgroup          Outfit
## Length:25           Min.    : 1      Min.    :0.6129
## Class :character    1st Qu.: 7      1st Qu.:0.6637
## Mode  :character    Median :13      Median :0.7756
##                     Mean    :13      Mean    :1.1011
##                     3rd Qu.:19      3rd Qu.:1.1267
```

```
##                            Max.    :25    Max.     :4.0039
##       Outfit_t                  Outfit_p
##    Min.    :-2.26286    Min.    :0.0000
##    1st Qu.:-1.38323    1st Qu.:0.1047
##    Median :-0.90572    Median :0.2016
##    Mean    : 0.03074    Mean    :0.3172
##    3rd Qu.: 0.41452    3rd Qu.:0.4431
##    Max.    : 7.73068    Max.    :0.9706
##       Infit                    Infit_t
##    Min.    :0.7045    Min.    :-2.2502
##    1st Qu.:0.8098    1st Qu.:-1.2284
##    Median :0.8953    Median :-0.7178
##    Mean    :1.0058    Mean    :-0.1539
##    3rd Qu.:1.0781    3rd Qu.: 0.5581
##    Max.    :2.1118    Max.     : 5.1137
##       Infit_p
##    Min.    :0.0000003
##    1st Qu.:0.1370592
##    Median :0.2876913
##    Mean    :0.3553087
##    3rd Qu.:0.5216361
##    Max.    :0.9243834
```

The `tam.fit()` function provides mean square error (MSE) and standardized (t) versions of the infit and outfit statistics for Rasch models. The `Infit` and `Outfit` statistics are the MSE versions, and the `Infit_t` and `Outfit_t` statistics are the standardized versions of the statistics. TAM also reports a p value for the standardized fit statistics (`Infit_p` and `Outfit_p`), along with adjusted significance values (`Infit_pholm` and `Outfit_pholm`). Please see Chapter 3 for a consideration of procedures for evaluating item fit in detail, including the use of graphical tools to evaluate item fit.

Person Parameters

Now we will examine person parameter estimates. With the MMLE procedure in TAM, person parameters are calculated after the item estimates using the `tam.wle()` function. The following code calculates person parameter estimates and saves them in an object called `person.locations_MMLE`.

```
# Use the tam.wle function to calculate person
    location parameters:
person.locations.estimate_MMLE <- tam.wle(PCM.science
    _MMLE)

## Iteration in WLE/MLE estimation    1    | Maximal
    change   1.1208
```

```
## Iteration in WLE/MLE estimation   2   | Maximal
      change  0.1558
## Iteration in WLE/MLE estimation   3   | Maximal
      change  0.0048
## Iteration in WLE/MLE estimation   4   | Maximal
      change  2e-04
## Iteration in WLE/MLE estimation   5   | Maximal
      change  0
## ----
##   WLE Reliability= 0.89
```

```
# Store person parameters and their standard errors
    in a dataframe object:
person.locations_MMLE <- cbind.data.frame(person.
    locations.estimate_MMLE$theta,

                                    person.locations
                                           .estimate_
                                           MMLE$error)
# Add descriptive column labels:
names(person.locations_MMLE) <- c("theta", "SE")

# View summary statistics for person parameters:
summary(person.locations_MMLE)
```

```
##       theta               SE
##  Min.   :-2.34123   Min.   :0.3329
##  1st Qu.:-0.70826   1st Qu.:0.3360
##  Median :-0.14167   Median :0.3468
##  Mean   : 0.07552   Mean   :0.3902
##  3rd Qu.: 0.74660   3rd Qu.:0.3775
##  Max.   : 4.61972   Max.   :1.4665
```

Because we centered the item location estimates at zero logits earlier, we need to adjust the person location estimates so that they can be compared to the centered item locations. We will do this by subtracting the original (uncentered) item mean from the person locations.

```
# Subtract the original (uncentered) item mean
    location from the person locations:
person.locations_MMLE$theta_adjusted <- person.
    locations_MMLE$theta - mean(uncentered.item.
    locations_MMLE)

# Summary of person location estimates:
summary(person.locations_MMLE)
```

```
##           theta                      SE
##    Min.    :-2.34123     Min.    :0.3329
##    1st Qu.:-0.70826     1st Qu.:0.3360
##    Median :-0.14167     Median :0.3468
##    Mean    : 0.07552     Mean    :0.3902
##    3rd Qu.: 0.74660     3rd Qu.:0.3775
##    Max.    : 4.61972     Max.    :1.4665
##    theta_adjusted
##    Min.    :-0.9218
##    1st Qu.: 0.7112
##    Median : 1.2778
##    Mean    : 1.4950
##    3rd Qu.: 2.1660
##    Max.    : 6.0392
```

If analysts do not prefer to use the zero-centered item location estimates, they should also use the original person parameters without subtracting the mean item location.

Person Fit

We can evaluate person fit using the `tam.personfit()` function from *TAM*. This function uses the model object as an argument and it produces outfit and infit statistics, as well as standardized versions of these statistics, for each person in the response matrix.

```
person.fit.results_MMLE <- tam.personfit(PCM.science_
   MMLE)
summary(person.fit.results_MMLE)
```

```
##    outfitPerson          outfitPerson_t
##    Min.    :0.01953     Min.    :-2.79344
##    1st Qu.:0.52351     1st Qu.:-0.85951
##    Median :0.71989     Median :-0.35538
##    Mean    :0.99717     Mean    : 0.08413
##    3rd Qu.:1.12701     3rd Qu.: 0.58252
##    Max.    :4.13636     Max.    : 4.75423
##    infitPerson          infitPerson_t
##    Min.    :0.04305     Min.    :-3.9199
##    1st Qu.:0.67419     1st Qu.:-1.1723
##    Median :0.84032     Median :-0.4058
##    Mean    :0.94655     Mean    :-0.2672
##    3rd Qu.:1.07093     3rd Qu.: 0.3322
##    Max.    :3.16633     Max.    : 5.1260
```

Wright Map

Next, we will create a Wright Map from our model results. This display will provide us with an overview of the distribution of person, item, and threshold parameters. We will create the plot by applying the *WrightMap* package function IRT.WrightMap() to the model object (PCM.science_MMLE).

For ease in interpretation, we will use the centered item and person locations that we calculated in this analysis. We need to specify these modified parameter estimates in the WrightMap() function. The following code prepares the parameter estimates and uses them to plot the Wright Map.

```
# Combine centered item estimates with thresholds:
n.items <- ncol(science.responses)

thresholds_MMLE <- matrix(data = NA, nrow = n.items,
    ncol = n.thresholds)

for(item.number in 1:n.items){
  for(tau in 1:n.thresholds){
  thresholds_MMLE[item.number, tau] <- centered.item.
     locations_MMLE[item.number] +
      tau.estimates_MMLE[item.number, tau]
  }
}

thetas_MMLE <- person.locations_MMLE$theta_adjusted

# Plot the Wright Map
wrightMap(thetas = thetas_MMLE,
         thresholds = thresholds_MMLE,
          main.title = "Liking for Science Partial
            Credit Model Wright Map (MMLE)",
         show.thr.lab   = TRUE, dim.names = "",
         label.items.rows= 2)
```

Liking for Science Partial Credit Model Wright Map (MMLE)

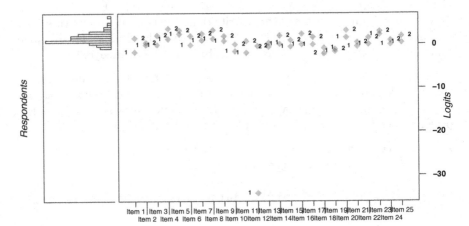

##		[,1]	[,2]
##	[1,]	-1.79168145	1.41355936
##	[2,]	-0.16413093	0.36483061
##	[3,]	-0.11920978	2.02924658
##	[4,]	1.19843634	3.54837428
##	[5,]	2.33835035	3.13800087
##	[6,]	-0.19918432	1.88895080
##	[7,]	0.87517628	2.40969400
##	[8,]	1.23863416	3.23105916
##	[9,]	0.70297360	1.84072361
##	[10,]	-1.63110913	-0.05567327
##	[11,]	-1.98196243	0.67059044
##	[12,]	-34.13678993	-0.43538289
##	[13,]	0.28619522	-0.63585593
##	[14,]	0.01635474	1.92692217
##	[15,]	-0.59136740	0.88589649
##	[16,]	-0.08132400	2.34874109
##	[17,]	-0.10901070	1.57702121
##	[18,]	-0.87840841	-2.14532771
##	[19,]	-1.65929441	-1.18369179
##	[20,]	1.53401531	3.13849189
##	[21,]	-0.54891527	0.36955709
##	[22,]	-0.05784055	1.42259340
##	[23,]	2.16136994	2.88987631
##	[24,]	-0.09798524	0.86263079
##	[25,]	0.33792095	1.85795848

In this *Wright Map* display, the results from the PCM analysis of the Liking for Science data are summarized graphically. The figure is organized as follows:

The left panel of the plot shows a histogram of respondent (children) locations on the logit scale that represents the latent variable. Units on the logit scale are shown on the far-right axis of the plot (labeled *Logits*).

The large central panel of the plot shows the rating scale category threshold estimates specific to each item on the logit scale that represents the latent variable. Light gray diamond shapes show the logit-scale location of the threshold estimates for each item, as labeled on the x-axis. Thresholds are labeled using an integer that shows the threshold number. In our example, τ_1 is the threshold between rating scale categories $x = 0$ and $x = 1$, and τ_2 is the threshold between rating scale categories $x = 1$ and $x = 2$.

Even though it is not appropriate to fully interpret item and person locations on the logit scale until there is evidence of acceptable model-data fit, we recommend examining the Wright Map during the preliminary stages of an item analysis to get a general sense of the model results and to identify any potential scoring or data entry errors.

A quick glance at the Wright Map suggests that, on average, the persons are located higher on the logit scale compared to the average item threshold locations. In addition, there appears to be a relatively wide spread of person and item locations on the logit scale, such that the Liking for Science questionnaire appears to be a useful tool for identifying differences in children's attitudes toward science activities as well as the difficulty to find each of the activities as favorable. Finally, the distance between rating scale category thresholds displays notable variability across the items in the survey. This finding supports the use of the PCM to analyze the Liking for Science data.

5.5   PCM Application with JMLE in *TAM*

In the following section, we provide an illustration of PCM analyses with the *Test Analysis Modules* or *TAM* package (Robitzsch et al., 2021) with JMLE. Except where there are significant differences between the MMLE and JMLE procedures, we provide fewer details about the analysis procedures and interpretations compared to the *eRm* and *TAM* MMLE illustrations.

Prepare for the Analyses

Please make sure that you have installed the *TAM* and *WrightMap* packages and loaded them into your working environment. You will also need to have imported the Liking for Science data and isolated the response matrix as described earlier in the chapter before you run the R code in the following sections.

Run the Partial Credit Model

We will use the `tam.jml()` function to run the PCM with JMLE and store the results in an object called `PCM.science_JMLE`. As with the MMLE procedure, in order to obtain Rasch-Andrich thresholds (i.e., adjacent-categories thresholds), we need to generate a design matrix for the model that includes specifications for those parameters. We will do this using the `designMatrices()` function from *TAM* and save the result in a new object called `design.matrix`.

```
design.matrix <- designMatrices(resp=science.
    responses, modeltype="PCM", constraint = "items")$
    A
```

Now we can run the PCM with our design matrix. After we run the model, we will request a summary of the model results using the `summary()` function.

```
PCM.science_JMLE <- tam.jml(science.responses, A =
    design.matrix, constraint = "items", control=list(
    maxiter=500), version=2 , verbose = FALSE)
```

```
summary(PCM.science_JMLE)
```

```
## -----------------------------------------------------
## TAM 3.7-16 (2021-06-24 14:31:37)
## R version 4.1.0 (2021-05-18) x86_64, darwin17.0 |
    nodename=XXXX | login=root
##
## Start of Analysis: 2021-10-27 13:46:48
## End of Analysis: 2021-10-27 13:46:48
## Time difference of 0.09117198 secs
## Computation time: 0.09117198
##
## Joint Maximum Likelihood Estimation in TAM
##
## IRT Model
## Call:
## tam.jml(resp = science.responses, A = design.
    matrix, constraint = "items",
##      verbose = FALSE, control = list(maxiter = 500)
    , version = 2)
##
## -----------------------------------------------------
## Number of iterations = 11
##
## Deviance = 2614.82  | Log Likelihood = -1307.41
## Number of persons = 75
## Number of items = 25
```

```
## constraint = items
## bias = TRUE
## -------------------------------------------------------
## Person Parameters xsi
## M = 0.19
## SD = 1.36
## -------------------------------------------------------
## Item Parameters xsi
##       item  N      M xsi.item AXsi_.Cat1 AXsi_.Cat2
## i1     i1 75  1.453   -1.504     -3.082     -3.008
## i2     i2 75  1.547   -1.235     -1.518     -2.469
## i3     i3 75  1.173   -0.380     -1.462     -0.761
## i4     i4 75  0.693    1.081     -0.170      2.161
## i5     i5 75  0.493    1.454      0.943      2.907
## i6     i6 75  1.213   -0.491     -1.540     -0.982
## i7     i7 75  0.920    0.307     -0.496      0.614
## i8     i8 75  0.720    0.927     -0.134      1.855
## i9     i9 75  1.067   -0.072     -0.668     -0.144
## i10   i10 75  1.733   -2.155     -2.941     -4.310
## i11   i11 75  1.613   -1.966     -3.271     -3.931
## i12   i12 75  1.827   -3.205     -4.651     -6.411
## i13   i13 75  1.693   -1.509     -1.094     -3.017
## i14   i14 75  1.173   -0.366     -1.332     -0.732
## i15   i15 75  1.480   -1.185     -1.927     -2.370
## i16   i16 75  1.107   -0.198     -1.423     -0.396
## i17   i17 75  1.267   -0.604     -1.455     -1.209
## i18   i18 75  1.933   -2.813     -2.230     -5.626
## i19   i19 75  1.880   -2.722     -2.977     -5.444
## i20   i20 75  0.667    1.029      0.153      2.058
## i21   i21 75  1.587   -1.420     -1.891     -2.840
## i22   i22 75  1.293   -0.657     -1.407     -1.314
## i23   i23 75  0.560    1.222      0.764      2.444
## i24   i24 75  1.427   -0.956     -1.449     -1.912
## i25   i25 75  1.133   -0.243     -1.021     -0.486
##      B.Cat1.Dim1 B.Cat2.Dim1
## i1             1           2
## i2             1           2
## i3             1           2
## i4             1           2
## i5             1           2
## i6             1           2
## i7             1           2
## i8             1           2
## i9             1           2
## i10            1           2
```

```
## i11              1            2
## i12              1            2
## i13              1            2
## i14              1            2
## i15              1            2
## i16              1            2
## i17              1            2
## i18              1            2
## i19              1            2
## i20              1            2
## i21              1            2
## i22              1            2
## i23              1            2
## i24              1            2
## i25              1            2
## -----------------------------------------------------
## Item Parameters  -A*Xsi
##     xsi.label xsi.index      xsi   se.xsi
## 1     i1_Cat1          1   -3.082   0.613
## 2     i1_Cat2          2    0.074   0.264
## 3     i2_Cat1          3   -1.518   0.417
## 4     i2_Cat2          4   -0.951   0.279
## 5     i3_Cat1          5   -1.462   0.337
## 6     i3_Cat2          6    0.701   0.287
## 7     i4_Cat1          7   -0.170   0.267
## 8     i4_Cat2          8    2.332   0.435
## 9     i5_Cat1          9    0.943   0.285
## 10    i5_Cat2         10    1.965   0.454
## 11    i6_Cat1         11   -1.540   0.347
## 12    i6_Cat2         12    0.559   0.282
## 13    i7_Cat1         13   -0.496   0.281
## 14    i7_Cat2         14    1.110   0.324
## 15    i8_Cat1         15   -0.134   0.268
## 16    i8_Cat2         16    1.989   0.401
## 17    i9_Cat1         17   -0.668   0.294
## 18    i9_Cat2         18    0.524   0.294
## 19   i10_Cat1         19   -2.941   0.749
## 20   i10_Cat2         20   -1.369   0.299
## 21   i11_Cat1         21   -3.271   0.742
## 22   i11_Cat2         22   -0.660   0.270
## 23   i12_Cat1         23   -4.651   1.845
## 24   i12_Cat2         24   -1.760   0.329
## 25   i13_Cat1         25   -1.094   0.452
## 26   i13_Cat2         26   -1.923   0.322
## 27   i14_Cat1         27   -1.332   0.329
```

```
## 28    i14_Cat2       28    0.600    0.286
## 29    i15_Cat1       29   -1.927    0.433
## 30    i15_Cat2       30   -0.443    0.269
## 31    i16_Cat1       31   -1.423    0.327
## 32    i16_Cat2       32    1.027    0.301
## 33    i17_Cat1       33   -1.455    0.349
## 34    i17_Cat2       34    0.246    0.275
## 35    i18_Cat1       35   -2.230    1.046
## 36    i18_Cat2       36   -3.396    0.540
## 37    i19_Cat1       37   -2.977    1.040
## 38    i19_Cat2       38   -2.466    0.399
## 39    i20_Cat1       39    0.153    0.269
## 40    i20_Cat2       40    1.906    0.405
## 41    i21_Cat1       41   -1.891    0.466
## 42    i21_Cat2       42   -0.949    0.279
## 43    i22_Cat1       43   -1.407    0.350
## 44    i22_Cat2       44    0.093    0.273
## 45    i23_Cat1       45    0.764    0.280
## 46    i23_Cat2       46    1.680    0.415
## 47    i24_Cat1       47   -1.449    0.379
## 48    i24_Cat2       48   -0.462    0.271
## 49    i25_Cat1       49   -1.021    0.311
## 50    i25_Cat2       50    0.534    0.288
```

Item Parameters

Next, we will examine the item difficulty location and rating scale category threshold estimates. As described earlier, we need to manually center the item locations at zero logits for ease of interpretation.

With the JMLE procedure, the overall item locations are stored in the xsi .item column of the the item table within the model object, which we will extract using the $ operator.

```
items_JMLE <- PCM.science_JMLE$item$xsi.item
```

Next, we will center the item parameter estimates that are stored in items_JMLE at zero logits, and request summary statistics for the estimates to check our work.

```
uncentered.item.locations_JMLE <- items_JMLE

centered.item.locations_JMLE <- scale(uncentered.item
    .locations_JMLE, scale = FALSE)
```

```
summary(centered.item.locations_JMLE)
```

```
##              V1
##    Min.   : -2.4989
##    1st Qu.: -0.7977
##    Median :  0.1020
##    Mean   :  0.0000
##    3rd Qu.:  0.6344
##    Max.   :  2.1602
```

We need to find the rating scale category threshold parameter estimates. The following code extracts the adjacent-categories threshold parameters from the item parameter table (item) that is stored in the PCM.science_JMLE object, and then stores them in an object called tau.estimates_JMLE.

```
# Specify the number of thresholds as the maximum
    observed score in the response matrix (be sure the
    responses begin at category 0):
n.thresholds <- max(science.responses)

# Save the threshold estimates, which begin in column
    5 of the item table:
tau.estimates_JMLE <- PCM.science_JMLE$item[, c(5 :
    (4+n.thresholds))]

# Specify the number of items that were included in
    the analysis:
n.items <- ncol(science.responses)

# Create a matrix in which to store the adjacent-
    category threshold values for each item:
tau.matrix_JMLE <- matrix(data = NA, ncol = n.
    thresholds, nrow = n.items)

# Calculate adjacent-category threshold values:

for(item.number in 1:n.items){
  for(tau in 1:n.thresholds){
    tau.matrix_JMLE[item.number, tau] <- ifelse(tau
        == 1,

                              (tau.estimates_JMLE
                                [item.number,
                                tau] -
                              uncentered.item.
                              locations_
```

```
                                            JMLE[item.
                                            number]),

                                        (tau.estimates_
                                            JMLE[item.
                                            number, tau]
                                            -
                                        sum(tau.
                                            estimates_
                                            JMLE[item.
                                            number, c
                                            ((tau-1))
                                            ]) -
                                        uncentered.
                                            item.
                                            locations_
                                            JMLE[item.
                                            number]))
        }
}

# View the threshold estimates:
tau.matrix_JMLE

##                  [,1]          [,2]
##   [1,]  -1.5779309   1.5779309
##   [2,]  -0.2833662   0.2833662
##   [3,]  -1.0816869   1.0816869
##   [4,]  -1.2511223   1.2511223
##   [5,]  -0.5111107   0.5111107
##   [6,]  -1.0495493   1.0495493
##   [7,]  -0.8029943   0.8029943
##   [8,]  -1.0617881   1.0617881
##   [9,]  -0.5957649   0.5957649
##  [10,]  -0.7861954   0.7861954
##  [11,]  -1.3058699   1.3058699
##  [12,]  -1.4455385   1.4455385
##  [13,]   0.4149299  -0.4149299
##  [14,]  -0.9655560   0.9655560
##  [15,]  -0.7423233   0.7423233
##  [16,]  -1.2249838   1.2249838
##  [17,]  -0.8508326   0.8508326
##  [18,]   0.5829667  -0.5829667
##  [19,]  -0.2552966   0.2552966
##  [20,]  -0.8763590   0.8763590
```

```
## [21,]  -0.4711518   0.4711518
## [22,]  -0.7497255   0.7497255
## [23,]  -0.4576070   0.4576070
## [24,]  -0.4935408   0.4935408
## [25,]  -0.7776856   0.7776856
```

Item Response Functions

Next, we will examine rating scale category probability plots. These plots have
the same format and interpretation as in the previous examples in this chapter.
To save space, we only printed the plots for the first five items here. Analysts
can request plots for different items using the `items` = argument in the `plot()`
function.

```
plot(PCM.science_JMLE, type="items", items = c(1:5))
```

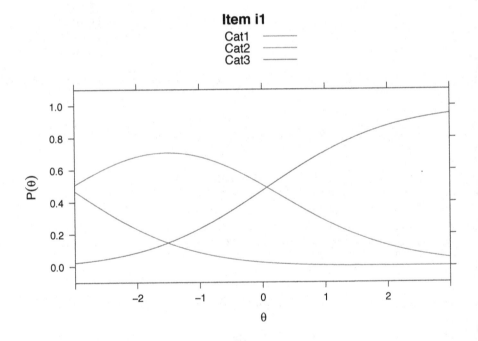

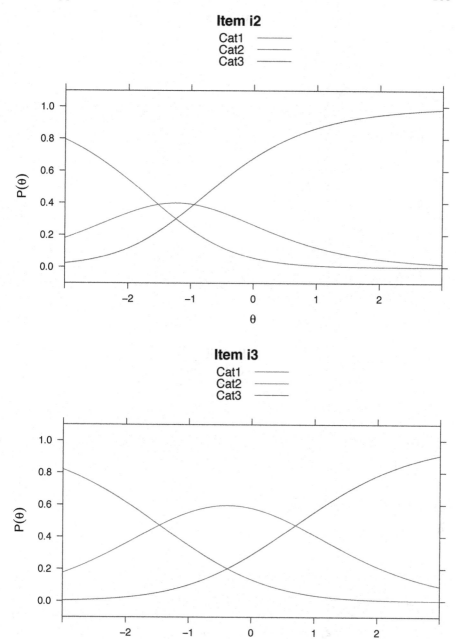

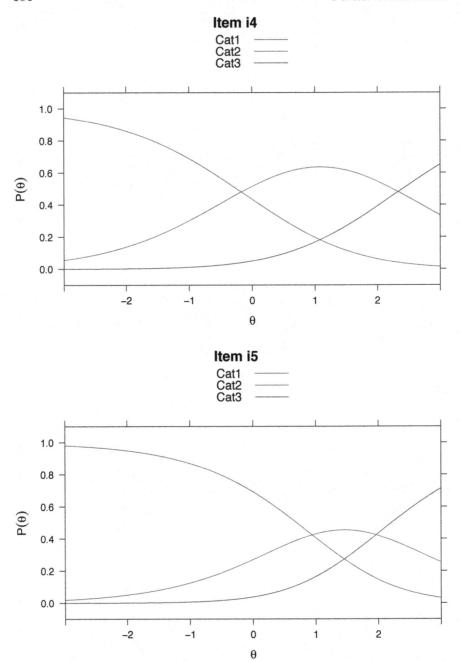

Item Fit

We will examine numeric item fit indices using the `itemfit()` function from *TAM*. We will save the results in an object called `item.fit_MMLE` and then view summary statistics for the fit statistics.

```
JMLE_fit <- tam.fit(PCM.science_JMLE)

item.fit_JMLE <- JMLE_fit$fit.item
summary(item.fit_JMLE)
```

```
##      item                 outfitItem
## Length:25            Min.    :0.5074
## Class :character     1st Qu.:0.5834
## Mode  :character     Median :0.7019
##                      Mean    :1.0086
##                      3rd Qu.:0.9848
##                      Max.    :3.7037
##   outfitItem_t          infitItem
## Min.    :-2.12203    Min.    :0.6388
## 1st Qu.:-1.17180    1st Qu.:0.7284
## Median :-0.86956    Median :0.8077
## Mean    : 0.02982    Mean    :0.9750
## 3rd Qu.: 0.16080    3rd Qu.:1.0729
## Max.    : 6.74340    Max.    :2.3418
##   infitItem_t
## Min.    :-2.8009
## 1st Qu.:-1.7114
## Median :-0.9834
## Mean    :-0.4069
## 3rd Qu.: 0.3603
## Max.    : 5.7831
```

The `tam.fit()` function provides mean square error (MSE) and standardized (t) versions of the outfit and infit statistics for Rasch models for each item estimate.

Person Parameters

Now we will examine person parameter estimates. With the JMLE procedure in *TAM*, person parameters are calculated at the same time as the item parameters. This means that we can extract the person parameter estimates from the model object without any additional iterations. The following code calculates person parameter estimates and saves them in an object called `person.locations_JMLE`.

```
# Store person parameters and their standard errors
    in a data.frame object:
```

```
person.locations_JMLE <- cbind.data.frame(PCM.science
    _JMLE$theta,
                                    PCM.science_JMLE
                                        $errorWLE)

names(person.locations_JMLE) <- c("theta", "SE")

# View summary statistics for person parameters:
summary(person.locations_JMLE)
```

```
##       theta                 SE
##   Min.    :-2.33409    Min.    :0.3366
##   1st Qu.:-0.65007    1st Qu.:0.3404
##   Median :-0.06078    Median :0.3529
##   Mean    : 0.18892    Mean    :0.3924
##   3rd Qu.: 0.87497    3rd Qu.:0.3857
##   Max.    : 5.31111    Max.    :1.1782
```

Because we centered the item location estimates at zero logits earlier, we need to adjust the person location estimates so that they can be compared to the centered item locations. We will do this by subtracting the original (uncentered) item mean from the person locations.

```
# Subtract the original (uncentered) item mean
    location from the person locations:
person.locations_JMLE$theta_adjusted <- person.
    locations_JMLE$theta - mean(uncentered.item.
    locations_JMLE)

# Summary of person location estimates:
summary(person.locations_JMLE)
```

```
##       theta                 SE
##   Min.    :-2.33409    Min.    :0.3366
##   1st Qu.:-0.65007    1st Qu.:0.3404
##   Median :-0.06078    Median :0.3529
##   Mean    : 0.18892    Mean    :0.3924
##   3rd Qu.: 0.87497    3rd Qu.:0.3857
##   Max.    : 5.31111    Max.    :1.1782
##   theta_adjusted
##   Min.    :-1.62762
##   1st Qu.: 0.05639
##   Median : 0.64569
##   Mean    : 0.89539
##   3rd Qu.: 1.58144
##   Max.    : 6.01758
```

If analysts do not prefer to use the zero-centered item location estimates, they can use the original person parameters without subtracting the mean item location.

Person Fit

Next, we will examine numeric person fit indices using the `tam.fit()` function from *TAM*. We will save the results in an object called `person.fit_JMLE` and then view summary statistics for the fit statistics.

```
JMLE_fit <- tam.fit(PCM.science_JMLE)

person.fit_JMLE <- JMLE_fit$fit.person

summary(person.fit_JMLE)
```

```
##     outfitPerson        outfitPerson_t
##   Min.    :0.01152    Min.     :-2.8598
##   1st Qu.:0.53832    1st Qu.:-0.8930
##   Median :0.72470    Median  :-0.3010
##   Mean    :1.00858    Mean     :  0.1177
##   3rd Qu.:1.15348    3rd Qu.:  0.5919
##   Max.    :4.16135    Max.     :  4.8677
##    infitPerson        infitPerson_t
##   Min.    :0.02696    Min.     :-3.9389
##   1st Qu.:0.68384    1st Qu.:-1.1288
##   Median :0.87020    Median  :-0.2847
##   Mean    :0.96830    Mean     :-0.2042
##   3rd Qu.:1.10773    3rd Qu.:  0.4384
##   Max.    :3.19143    Max.     :  5.1642
```

The `tam.fit()` function provides mean square error (MSE) and standardized (t) versions of the outfit and infit statistics for each person location estimate.

Wright Map

Finally, we will create a Wright Map from our model results. This display will provide us with an overview of the distribution of person, item, and threshold parameters. We will create the plot by applying the *WrightMap* package function `IRT.WrightMap()` to the model object (`PCM.science_JMLE`).

For ease in interpretation, we will use the centered item and person locations that we calculated in this analysis. We need to specify these modified parameter estimates in the `WrightMap()` function. The following code prepares the parameter estimates and plots the Wright Map using them.

```
# Combine centered item estimates with thresholds:
n.items <- ncol(science.responses)
```

```
thresholds_JMLE <- matrix(data = NA, nrow = n.items,
    ncol = n.thresholds)

for(item.number in 1:n.items){
  for(tau in 1:n.thresholds){
  thresholds_JMLE[item.number, tau] <- centered.item.
    locations_JMLE[item.number] +
      tau.matrix_JMLE[item.number, tau]
  }
}


thetas_JMLE <- person.locations_JMLE$theta_adjusted

# Plot the Wright Map
wrightMap(thetas = thetas_JMLE,
          thresholds = thresholds_JMLE,
          main.title = "Liking for Science Partial
              Credit Model Wright Map (JMLE)",
          show.thr.lab    = TRUE, dim.names = "",
          label.items.rows= 2)
```

Liking for Science Partial Credit Model Wright Map (JMLE)

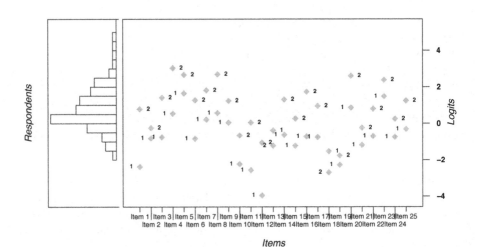

```
##                 [,1]          [,2]
## [1,]    -2.37565811    0.78020364
## [2,]    -0.81154866   -0.24481629
## [3,]    -0.75565594    1.40771779
## [4,]     0.53609201    3.03833660
## [5,]     1.64910466    2.67132610
```

```
##   [6,]  -0.83401562   1.26508307
##   [7,]   0.21026536   1.81625405
##   [8,]   0.57198615   2.69556237
##   [9,]   0.03864129   1.23017104
##  [10,]  -2.23484689  -0.66245618
##  [11,]  -2.56491729   0.04682261
##  [12,]  -3.94441860  -1.05334159
##  [13,]  -0.38715847  -1.21701823
##  [14,]  -0.62514270   1.30596930
##  [15,]  -1.22085899   0.26378767
##  [16,]  -0.71674590   1.73322169
##  [17,]  -0.74880318   0.95286201
##  [18,]  -1.52356629  -2.68949970
##  [19,]  -2.27058296  -1.75998978
##  [20,]   0.85934916   2.61206726
##  [21,]  -1.18481711  -0.24251350
##  [22,]  -0.70027394   0.79917716
##  [23,]   1.47088270   2.38609670
##  [24,]  -0.74296829   0.24411333
##  [25,]  -0.31442532   1.24094583
```

5.6   Example Results Section

```
# Print Table 1:
tab1 <- knitr::kable(
  PCM_Table1, booktabs = TRUE,
  caption = 'Model Summary Table'
)
tab1 %>%
  kable_styling(latex_options = "scale_down", full_
    width = FALSE)
```

Table 5.1 presents a summary of the results from the analysis of the Liking for Science data (Wright and Masters, 1982) using the PCM (Masters, 1982). One child was excluded from the analysis (Child #2), who gave extreme responses to all of the items ($x = 2$).

Specifically, Table 5.1 summarizes the calibration of the children with non-extreme responses ($N = 74$) and the items ($N = 25$) using average logit-scale calibrations, standard errors, and model-data fit statistics. Examination of the results indicates that, on average, the children were located higher on the logit scale ($M = 0.96$, $SD = 1.19$), compared to items, whose locations were

TABLE 5.1

Model Summary Table

Statistic	Items	Persons
Logit Scale Location Mean	0.00	0.96
Logit Scale Location SD	1.20	1.19
Standard Error Mean	0.40	0.38
Standard Error SD	0.18	0.10
Outfit MSE Mean	1.06	1.06
Outfit MSE SD	0.94	1.00
Infit MSE Mean	0.97	0.97
Infit MSE SD	0.42	0.48
Std. Outfit Mean	−0.14	0.15
Std. Outfit SD	2.63	1.72
Std. Infit Mean	−0.44	−0.23
Std. Infit SD	2.08	1.55
Separation.reliability	1.00	0.89

centered at zero logits ($M = 0.00$, $SD = 1.20$). This finding suggests that the items were relatively easy to endorse among the sample of children who participated in this study. The average values of the Standard Error (SE) were comparable for children ($M = 0.38$) and items ($M = 0.40$). Average values of model-data fit statistics indicate overall adequate fit to the model, with average outfit and infit mean square error (MSE) statistics around 1.00, and average standardized outfit and infit statistics near the expected value of 0.00 when data fit the model. However, there was some variability in item fit and person fit, as indicated by relatively large standard deviations for the fit statistics. Additional investigation into item fit and person fit is warranted.

```
# Print Table 2:
tab2 <- knitr::kable(
  PCM_Table2, booktabs = TRUE,
  caption = 'Item Calibrations'
)
tab2 %>%
  kable_styling(latex_options = "scale_down", full_
      width = FALSE)
```

Table 5.2 includes detailed results for the 25 Liking for Science items, where items are ordered by their overall logit-scale location (i.e., item difficulty) from high (difficult to endorse) to low (easy to endorse). For each item, the average rating is presented, followed by the overall logit-scale location (δ), the location of the item-specific rating scale category thresholds, and item fit statistics.

TABLE 5.2

Item Calibrations

	Task ID	Average Rating	Item Location	Threshold 1	Threshold 2	MSE Outfit	Std. Outfit	Infit MSE	Std. Infit
i5	5	0.49	2.08	−0.46	0.46	3.72	6.65	2.20	5.05
i23	23	0.56	1.85	−0.41	0.41	4.36	8.17	2.28	5.58
i4	4	0.69	1.72	−1.23	1.23	0.95	−0.28	0.96	−0.26
i20	20	0.67	1.66	−0.84	0.84	1.71	3.39	1.31	1.89
i8	8	0.72	1.56	−1.03	1.03	1.19	1.16	1.13	0.90
i7	7	0.92	0.95	−0.78	0.78	0.98	−0.10	0.94	−0.38
i9	9	1.07	0.57	−0.56	0.56	1.10	0.61	1.08	0.63
i16	16	1.11	0.45	−1.21	1.21	1.02	0.17	1.06	0.44
i25	25	1.13	0.40	−0.75	0.75	0.70	−1.81	0.76	−1.81
i14	14	1.17	0.28	−0.95	0.95	0.78	−1.30	0.85	−1.07
i3	3	1.17	0.27	−1.07	1.07	0.61	−2.67	0.64	−2.83
i6	6	1.21	0.16	−1.03	1.03	0.80	−1.17	0.86	−0.99
i17	17	1.27	0.05	−0.83	0.83	0.59	−2.49	0.65	−2.78
i22	22	1.29	−0.01	−0.72	0.72	0.70	−1.61	0.78	−1.58
i24	24	1.43	−0.30	−0.46	0.46	0.67	−1.34	0.77	−1.60
i15	15	1.48	−0.53	−0.72	0.72	0.59	−1.81	0.73	−1.85
i2	2	1.55	−0.58	−0.24	0.24	0.54	−1.55	0.73	−1.70
i21	21	1.59	−0.76	−0.43	0.43	0.53	−1.63	0.71	−1.78
i13	13	1.69	−0.84	0.48	−0.48	0.66	−0.52	0.80	−0.87
i1	1	1.45	−0.86	−1.58	1.58	0.65	−1.96	0.72	−2.06
i12	12	1.83	−1.12	0.00	NA	0.57	−0.98	0.81	−1.05
i11	11	1.61	−1.32	−1.30	1.30	0.58	−1.78	0.71	−1.94
i10	10	1.73	−1.50	−0.75	0.75	0.54	−1.27	0.75	−1.26
i19	19	1.88	−2.05	−0.19	0.19	1.02	0.26	0.91	−0.16
i18	18	1.93	−2.13	0.68	−0.68	1.04	0.43	1.12	0.40

Item 5 was the most difficult to endorse (*AverageRating* = 0.49; δ = 2.08), followed by Item 23 (*AverageRating* = 0.56; δ = 1.85). The easiest item to endorse was Item 18 (*AverageRating* = 1.93; δ = −2.13).

```
# Print Table 3:
tab3 <- knitr::kable(
  head(PCM_Table3,10), booktabs = TRUE,
  caption = 'Person Calibration'
)
tab3 %>%
  kable_styling(latex_options = "scale_down", full_
    width = FALSE)
```

TABLE 5.3
Person Calibration

Child ID	Average Rating	Person Location	Person SE	Outfit MSE	Std. Outfit	Infit MSE	Std. Infit
1	1.16	0.67	0.34	0.82	−0.39	0.93	−0.18
3	1.32	1.16	0.36	0.41	−1.52	0.45	−2.39
4	1.04	0.33	0.33	0.64	−1.11	0.70	−1.18
5	0.72	−0.55	0.34	1.32	1.01	0.78	−0.85
6	0.92	0.00	0.33	2.36	3.22	1.54	1.90
7	1.72	2.74	0.47	1.16	0.47	1.81	1.86
8	1.28	1.03	0.35	0.83	−0.26	0.99	0.05
9	0.92	0.00	0.33	1.89	2.34	1.37	1.39
10	1.08	0.44	0.34	0.64	−1.05	0.71	−1.12
11	1.36	1.28	0.36	0.79	−0.30	0.99	0.05

Table 5.3 includes detailed results for 10 children who participated in the Liking for Science survey. For each child, the average rating is presented, followed by their logit-scale location estimate (θ), SE, and model-data fit statistics.

```
plotPImap(PCM.science, main = "Liking for Science
    Partial Credit Model Wright Map", sorted = TRUE,
    irug = FALSE)
```

Liking for Science Partial Credit Model Wright Map

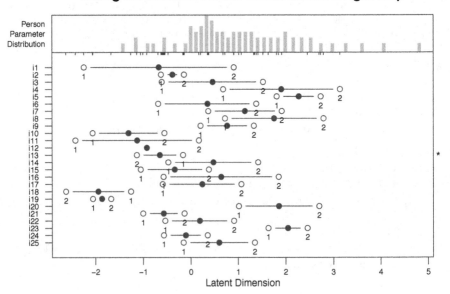

Figure 1 illustrates the calibrations of the children and items on the logit scale that represents the latent variable. The calibrations in this figure correspond to the results presented in Tables 5.2 and 5.3 for items and children, respectively. Starting at the bottom of the figure, the horizontal axis (labeled *Latent Dimension*) is the logit scale that represents the latent variable. Lower numbers on this scale indicate less-favorable attitudes toward science activities, and higher numbers indicate more-favorable attitudes toward science activities. The central panel of the figure shows item difficulty locations on the logit scale for the 25 Liking for Science items included in the analysis; the y-axis for this panel shows the item labels. The items are ordered according to their difficulty order, as estimated with the PCM. Easier-to-endorse items appear at the top of the figure, and harder-to-endorse items appear at the bottom of the figure; item labels are shown on the y-axis. For each item, a solid circle plotting symbol shows the overall location estimate. This solid circle symbol is connected to two open-circle symbols that show the locations of the rating scale category thresholds. Each threshold is labeled with a 1 to indicate the threshold between rating scale category $x = 0$ and $x = 1$, or a 2 to indicate the threshold between rating scale category $x = 1$ and $x = 2$. The threshold locations were ordered as expected for most items, with τ_1 located lower on the logit scale (τ) compared to the location estimate for τ_2. However, the thresholds were disordered for Item 13 and Item 18. In addition, there were some differences in the relative difficulty to endorse the rating scale categories across items. Finally, the upper panel of the figure shows a histogram of person (in this case, children) location estimates on the logit scale.

5.7 Exercise

Please use the *eRm* or *TAM* package to estimate item, threshold, and person locations with the PCM for the Exercise 5 data. The Exercise 5 data include responses from 500 participants to an attitude survey that includes 20 items with a 5-category rating scale. The Chapter 5 exercise data is the same as the Exercise 4 data. After you complete the analysis, try writing a results section similar to the example in this chapter to report on your findings.

6

Many Facet Rasch Model

This chapter provides a basic overview of the Many-Facet Rasch Model (MFRM) (Linacre, 1989), along with guidance for analyzing data with the MFRM using R. We use the *TAM* package (Robitzsch et al., 2021) for all of the analyses in this chapter. In the first example, we demonstrate how to apply the MFRM to multi-faceted data that are stored in *wide format* (one row for each subject). Then, we demonstrate how to apply the MFRM to multi-faceted data that are stored in *long format* (multiple rows for each subject). After the analyses are complete, we present an example description of the results from one example. The chapter concludes with a challenge exercise.

Overview of the Many-Facet Rasch Model

The MFRM (Linacre, 1989) is an extension of the Rasch family of models that allows researchers to include variables of interest ("facets") besides items and persons. Bond et al. (2019) defined facets as aspects of the measurement process that "routinely and systematically interpose themselves between the ability of the candidates and the difficulty of the test" (p. 145). Examples of variables that could be modeled as facets include raters in a constructed-response assessment, participant demographic variables (e.g., gender, race/ethnicity, best language), item type, or domains in an analytic scoring rubric for constructed-response items.

A general equation for the MFRM is:

$$\ln \left(\frac{P_{n(x=k)}}{P_{n(x=k-1)}} \right) = \theta_n - \sum_{\text{facets}} \varepsilon - \tau_k \qquad (6.1)$$

In Equation 6.1, θ is the person location parameter and τ_k is the rating scale category threshold, which can be modified to reflect different variations on the Rasch model (e.g., the PC model) as needed. Σ facets ϵ is a linear combination of facets that are specific to each modeling context. For example, facets for a MFRM analysis of a performance assessment could include raters and domains. According to Equation 6.1, the probability for an observation in category k, rather than in category $k-1$ for Person n is modeled as the difference between the location of Person n, the location of the researcher-specified facets, and the difficulty associated with providing a response in category k.

Researchers can specify formulations of the MFR model to extend the dichotomous Rasch model (see Chapter 2) the Rating Scale (RS) model (see Chapter
4), the Partial Credit model (see Chapter 5), as well as other Rasch models
(e.g., the binomial trials model and the Poisson counts model (Wright and
Mok, 2004).

Rasch Model Requirements

The MFRM shares the requirements for unidimensionality, local independence,
and invariance that we discussed in Chapter 2 for the dichotomous Rasch
model. In practice, researchers should evaluate item responses for evidence
that they approximate Rasch model requirements before examining model
estimates in detail. Chapter 3 included details about model-data fit analysis
procedures that can also be applied to the MFRM.

6.1   Running the MFRM with Wide-Format Data in TAM Package

In the next section, we provide a step-by-step demonstration of a MFRM
analysis using the *Test Analysis Modules* or *TAM* package (Robitzsch et al.,
2021) for R. We encourage readers to use the example data set for this chapter
that is provided in the online supplement to conduct the analysis along with
us.

For this first example, we use a subset of writing assessment data that includes
students' scores related to the style of their writing. In the second example in
this chapter, we use students' scores related to four domains.

Prepare for the Analyses

Use the following code to get started with the *TAM* package by installing it
and loading it into your R environment.

```
# install.packages("TAM")
library("TAM")
```

We will also use the *WrightMap* package (Irribarra and Freund, 2014).

```
# install.packages("WrightMap")
library("WrightMap")
```

Finally, we will use the *psych* package (Revelle, 2021).

```
# install.packages("psych")
library("psych")
```

Now that we have installed and loaded the packages to our R session, we are ready to import the data. We will use the function `read.csv()` to import the comma-separated values (.csv) file that contains the data for the first example. We encourage readers to use their preferred method for importing data files into R or R Studio.

Please note that if you use `read.csv()` to import the data, you will first need to specify the file path to the location at which the data file is stored on your computer or set your working directory to the folder in which you have saved the data.

First, we will import the data using `read.csv()` and store it in an object called `style`.

```
style <- read.csv("style_ratings.csv")
```

The style ratings file is in *wide format*, because there is one row for each of the 372 unique students. We can see this structure by printing the first six rows of the data frame object to our console using the `head()` function.

```
head(style)
```

```
##      student language rater_1 rater_2 rater_3
## 1          3        1       2       2       2
## 2          7        1       2       3       2
## 3         11        1       2       2       2
## 4         15        1       0       0       0
## 5         20        1       1       1       1
## 6         24        2       0       0       0
##      rater_4 rater_5 rater_6 rater_7 rater_8 rater_9
## 1          2       3       3       1       2       2
## 2          3       1       1       1       2       1
## 3          2       2       3       2       2       2
## 4          1       0       0       1       0       0
## 5          2       2       1       1       1       1
## 6          0       0       0       0       0       0
##      rater_10 rater_11 rater_12 rater_13 rater_14
## 1           2        2        2        1        1
## 2           2        3        3        2        1
## 3           2        2        2        2        2
## 4           0        0        1        1        0
## 5           1        1        2        1        0
## 6           0        0        0        0        0
##      rater_15 rater_16 rater_17 rater_18 rater_19
## 1           2        2        2        1        1
## 2           1        2        1        2        2
## 3           2        2        2        2        2
```

```
## 4           0         0        1        0        0
## 5           1         1        1        1        1
## 6           0         0        1        0        0
##    rater_20 rater_21
## 1         2        1
## 2         1        2
## 3         2        1
## 4         0        0
## 5         1        2
## 6         0        0
```

Next, we will explore the data using descriptive statistics with the summary()
function.

summary(style)

```
##     student           language          rater_1
##  Min.   :   3.0    Min.   :1.000    Min.   :0.00
##  1st Qu.: 392.0    1st Qu.:1.000    1st Qu.:1.00
##  Median : 782.0    Median :2.000    Median :2.00
##  Mean   : 784.3    Mean   :1.532    Mean   :1.68
##  3rd Qu.:1171.2    3rd Qu.:2.000    3rd Qu.:2.00
##  Max.   :1574.0    Max.   :2.000    Max.   :3.00
##     rater_2           rater_3          rater_4
##  Min.   :0.000    Min.   :0.000    Min.   :0.000
##  1st Qu.:1.000    1st Qu.:1.000    1st Qu.:1.000
##  Median :2.000    Median :2.000    Median :2.000
##  Mean   :1.688    Mean   :1.573    Mean   :1.605
##  3rd Qu.:2.000    3rd Qu.:2.000    3rd Qu.:2.000
##  Max.   :3.000    Max.   :3.000    Max.   :3.000
##     rater_5           rater_6          rater_7
##  Min.   :0.000    Min.   :0.000    Min.   :0.000
##  1st Qu.:1.000    1st Qu.:1.000    1st Qu.:1.000
##  Median :2.000    Median :2.000    Median :2.000
##  Mean   :1.798    Mean   :1.694    Mean   :1.667
##  3rd Qu.:3.000    3rd Qu.:2.000    3rd Qu.:2.000
##  Max.   :3.000    Max.   :3.000    Max.   :3.000
##     rater_8           rater_9          rater_10
##  Min.   :0.000    Min.   :0.000    Min.   :0.00
##  1st Qu.:1.000    1st Qu.:1.000    1st Qu.:1.00
##  Median :2.000    Median :1.000    Median :2.00
##  Mean   :1.548    Mean   :1.347    Mean   :1.82
##  3rd Qu.:2.000    3rd Qu.:2.000    3rd Qu.:3.00
##  Max.   :3.000    Max.   :3.000    Max.   :3.00
##     rater_11          rater_12         rater_13
##  Min.   :0.000    Min.   :0.00     Min.   :0.000
```

```
## 1st Qu.:1.000      1st Qu.:1.00      1st Qu.:1.000
## Median :2.000      Median :2.00      Median :2.000
## Mean   :1.664      Mean   :1.68      Mean   :1.694
## 3rd Qu.:2.000      3rd Qu.:2.00      3rd Qu.:2.000
## Max.   :3.000      Max.   :3.00      Max.   :3.000
##    rater_14            rater_15            rater_16
## Min.   :0.000      Min.   :0.000      Min.   :0.000
## 1st Qu.:1.000      1st Qu.:1.000      1st Qu.:1.000
## Median :2.000      Median :2.000      Median :1.000
## Mean   :1.551      Mean   :1.809      Mean   :1.508
## 3rd Qu.:2.000      3rd Qu.:2.000      3rd Qu.:2.000
## Max.   :3.000      Max.   :3.000      Max.   :3.000
##    rater_17            rater_18            rater_19
## Min.   :0.000      Min.   :0.000      Min.   :0.000
## 1st Qu.:1.000      1st Qu.:1.000      1st Qu.:1.000
## Median :2.000      Median :2.000      Median :1.000
## Mean   :1.505      Mean   :1.586      Mean   :1.454
## 3rd Qu.:2.000      3rd Qu.:2.000      3rd Qu.:2.000
## Max.   :3.000      Max.   :3.000      Max.   :3.000
##    rater_20            rater_21
## Min.   :0.000      Min.   :0.000
## 1st Qu.:1.000      1st Qu.:1.000
## Median :2.000      Median :2.000
## Mean   :1.508      Mean   :1.492
## 3rd Qu.:2.000      3rd Qu.:2.000
## Max.   :3.000      Max.   :3.000
```

From the summary of style, we can see there are no missing data. In addition, we can get a general sense of the scales, range, and distribution of each variable in the data set. For example, we can see that the data include student identification numbers, a language subgroup variable, and ratings from 21 raters. Student identification numbers range from 3 to 1574. There are two language subgroups: Subgroup 1 (*language* = 1) indicates that students' best language is a language other than English, and subgroup 2 (*language* = 2) indicates that students' best language is English. The minimum rating from each rater was $x = 0$, and the maximum rating was $x = 3$.

Please note that the *TAM* package requires that the lowest observation for item responses is equal to zero. In our data, this property is already present. If the lowest category is something other than zero, the analyst will need to re-code the responses as we have done in previous chapters.

Specify Model Components

Now, we are ready to run the MFRM on the style ratings. Because the MFRM equation is researcher-specified, we need to define the components of the model. To do this, we will create an object called facets in which we specify the facets in the model. By default, the *TAM* package treats the variables that make up the columns of our item response matrix as an "item" facet. In our example, raters function as pseudo-items. Accordingly, raters make up the first facet in our analysis. Our second facet will be student language subgroups. We specify this facet and save it in a data frame object called `facets`. We specified `drop = FALSE` because the data frame only includes one column.

```
facets <- style[ , "language", drop = FALSE]
```

Next we need to identify the indicator variable for the object of measurement (i.e., subject). In our example, students are the object of measurement. We will save the student identification numbers in a vector called `students`.

```
students <- style$student
```

Finally, we need to specify the response matrix. We do so by extracting the raters' scores for each student to a data frame object called `ratings`.

```
ratings <- subset(style, select = -c(student,
    language))
```

Next, we need to specify the formula for our MFRM. For the first example, we will use a rating scale model specification of the MFRM. This means that we will constrain the threshold parameters to be equal across raters. The model is specified as follows:

$$\ln \left(\frac{P_{nji(x=k)}}{P_{nji(x=k-1)}} \right) = \theta_n - \gamma_j - \lambda_i - \tau_k \tag{6.2}$$

In Equation 6.2, θ_n is defined as in Equation 6.1. γ_j is the logit-scale location for student language subgroup j, λ_i is the logit-scale location for rater i, and τ_k is the logit-scale location at which there is an equal probability for a rating in category k and category $k - 1$. The subgroup facet (γ_j) reflects the overall location of students in subgroup j, where higher locations indicate higher levels of writing proficiency, and lower locations indicate lower levels of writing proficiency. The rater facet (λ_i) reflects the severity level of individual rater i, where higher locations mean that the rater is more severe, and requires higher levels of writing proficiency before giving high ratings to student performances. Lower rater locations indicate relatively lenient raters, who readily give high ratings to student performances.

We specify the MFRM from Equation 6.2 in an object for use with *TAM* as follows. First, we specify a name for the model object (`style_RS_MFRM`), which is defined using the tilde symbol (~), followed by the facet names. As a reminder, the model must include a facet named *item*; in our example, the item facet is made up of raters. We also include the student language subgroup (*language*) as a facet. Finally, we specify *step* to indicate the RS model. The components of the model are separated by addition signs (+) because the facets are additive. We will use interactions in a MFRM later in this chapter.

```
style_RS_MFRM <- ~ item + language + step
```

Run the RS-MFRM

Now we can run our RS-MFRM. We do so using the tam.mml.mfr() function, in which we specify the response matrix (resp =), our specified facets (facets =), the model equation (formulaA =), and the identification numbers for the object of measurement (pid =).

```
RS_MFR_model <- tam.mml.mfr(resp = ratings, facets =
    facets, formulaA = style_RS_MFRM, pid = students,
    constraint = "items", verbose = FALSE)
```

Overall Model Summary

After we run the model, we will request a summary of the model results using the summary() function.

```
summary(RS_MFR_model)
```

The summary of the MFRM is lengthy, and we do not reproduce it here. Included among the output are several details that may be important for some analyses, including details about each iteration, global model-fit indicators (e.g., deviance, log-likelihood, AIC, BIC) and an estimate of reliability. We will focus our interpretation on the location estimates for the student, subgroup, rater, and threshold parameters.

Facet Results

Next, we will save the parameter estimates from the RS-MFRM in a data frame object called facet.estimates. This object includes the location estimates and standard errors for raters, student subgroups, and thresholds. Location estimates are labeled xsi and standard errors are labeled se.xsi.

```
facet.estimates <- RS_MFR_model$xsi.facets
```

For easier reference, we will now create objects in which we store the location estimates and standard errors for raters, subgroups, and thresholds separately. We do this by applying the subset() function to the facet.estimates object.

```
rater.estimates <- subset(facet.estimates, facet.
    estimates$facet == "item")

subgroup.estimates <- subset(facet.estimates, facet.
    estimates$facet == "language")

threshold.estimates <- subset(facet.estimates, facet.
    estimates$facet == "step")
```

Rater Facet Results

Next, we will request a summary of the rater facet estimates.

```
summary(rater.estimates)
```

```
##    parameter                  facet
## Length:21              Length:21
## Class  :character      Class  :character
## Mode   :character      Mode   :character
##
##
##
##         xsi                      se.xsi
## Min.    :-0.86778      Min.    :0.07494
## 1st Qu.:-0.31386      1st Qu.:0.07497
## Median : 0.03373      Median :0.07499
## Mean    : 0.00000      Mean    :0.08744
## 3rd Qu.: 0.43770      3rd Qu.:0.07509
## Max.    : 1.11732      Max.    :0.33558
```

From the summary of rater locations, we can see that rater severity estimates range from $\lambda = -0.87$ for the most lenient rater to $\lambda = 1.12$ for the most severe rater. The average rater severity location is set to zero logits.

Student Subgroup Facet Results

Next, we will examine the language subgroup estimates using the summary() function.

```
summary(subgroup.estimates$xsi)
```

```
##      Min.   1st Qu.   Median      Mean   3rd Qu.
## -0.06565  -0.03283  0.00000  0.00000  0.03283
##      Max.
##  0.06565
```

We can also print the locations to our console to inspect them.

```
subgroup.estimates$xsi
```

```
## [1]  -0.06565052    0.06565052
```

We can see that the two student subgroup locations are quite close to one another. As a group, students whose best language is a language other than English (*language* = 1) had a slightly lower location on the logit scale ($\gamma_1 = -0.07$ logits) compared to students whose best language was English (*language* 2; $\gamma_2 = 0.07$ logits). Although there was a difference in subgroup locations, the difference was very small (about 0.13 logits), and therefore likely does not reflect a substantively meaningful difference in writing achievement between these two groups.

Student estimates

Next, we will examine the student location estimates from the RS-MFRM. We will apply the `tam.wle()` function to our model object (`RS_MFR_model`) and store the results in an object called `student.ach`. Then, we will store the student identification numbers, location estimates, and standard errors in a new object called `student.locations_RSMFR`. Finally, we will summarize the results using the `summary()` function.

```
student.ach <- tam.wle(RS_MFR_model)

student.locations_RSMFR <- cbind.data.frame(student.
    ach$pid, student.ach$theta, student.ach$error)

names(student.locations_RSMFR) <- c("id", "theta", "
    se")

summary(student.locations_RSMFR)
```

From the summary of the student achievement locations, we can see that student achievement ranges from $\theta = -7.96$ logits for the student with the lowest achievement estimate to $\theta = 7.60$ for the student with the highest achievement estimate. On average, the students were located slightly higher $(M\ \theta) = 0.47)$ than the average rater location $(M\ \lambda = 0.00)$.

Threshold Estimates

Next we will examine the threshold estimates. Because we used a RS formulation of the MFRM, there is one set of threshold estimates for our model that applies across raters. We will print the threshold estimates to the console to view them.

```
threshold.estimates$xsi

## [1] -4.0087904  0.2353612  3.7734293
```

As we discussed in Chapters 4 and 5, analysts can evaluate the order of the threshold locations for evidence that they are non-decreasing across increasing rating scale categories.

Wright Map

Next, we will plot a Wright Map to display the locations of the parameter estimates for our RS-MFR model. To do this, we need to manipulate the location estimate objects into the format required for the *WrightMap* package.

First, we need to store the rater location estimates as a matrix that shows rater-specific threshold locations. We accomplish this task using a for-loop in which we add each rater's location to the three threshold values and store the results in a matrix called `rater_thresholds`. We use the `head()` function to preview the first six rows of the results in the console.

```
rater_thresholds <- matrix(data = NA, nrow = nrow(
    rater.estimates), ncol = nrow(threshold.estimates)
    )

for(rater in 1:nrow(rater.estimates)){
  for(tau in 1:nrow(threshold.estimates)){
  rater_thresholds[rater,tau] <- (rater.estimates$xsi
      [rater] + threshold.estimates$xsi[tau])
  }
}

head(rater_thresholds)
```

```
##               [,1]          [,2]        [,3]
## [1,]      -4.288967  -0.04481525  3.493253
## [2,]      -4.322646  -0.07849438  3.459574
## [3,]      -3.840553   0.40359889  3.941667
## [4,]      -3.975059   0.26909229  3.807160
## [5,]      -4.785429  -0.54127709  2.996791
## [6,]      -4.345107  -0.10095584  3.437112
```

Finally, we can plot the Wright map using the `wrightMap()` function. We specify several graphical parameters to modify the appearance of the plot.

```
wrightMap(thetas = cbind(student.locations_RSMFR$
    theta, subgroup.estimates$xsi),
          axis.persons = "Students",
          dim.names = c("Students", "Subgroups"),
          thresholds = rater_thresholds,
          show.thr.lab  = TRUE,
          label.items.rows= 2,
          label.items = rater.estimates$parameter,
          axis.items = "Raters",
          main.title = "Rating Scale Many-Facet Rasch
              Model Wright Map:\n Style Ratings",
          cex.main = .5)
```

Rating Scale Many–Facet Rasch Model Wright Map: Style Ratings

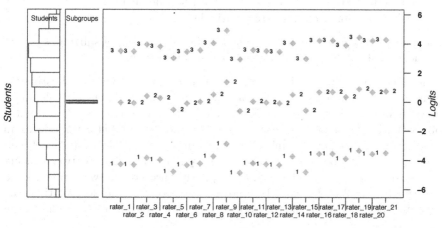

```
##              [,1]          [,2]         [,3]
##   [1,]   -4.288967   -0.04481525   3.493253
##   [2,]   -4.322646   -0.07849438   3.459574
##   [3,]   -3.840553    0.40359889   3.941667
##   [4,]   -3.975059    0.26909229   3.807160
##   [5,]   -4.785429   -0.54127709   2.996791
##   [6,]   -4.345107   -0.10095584   3.437112
##   [7,]   -4.232865    0.01128652   3.549355
##   [8,]   -3.739598    0.50455410   4.042622
##   [9,]   -2.891467    1.35268451   4.890753
##  [10,]   -4.876566   -0.63241444   2.905654
##  [11,]   -4.221649    0.02250299   3.560571
##  [12,]   -4.288967   -0.04481525   3.493253
##  [13,]   -4.345107   -0.10095584   3.437112
##  [14,]   -3.750819    0.49333235   4.031400
##  [15,]   -4.830952   -0.58680019   2.951268
##  [16,]   -3.571092    0.67305934   4.211127
##  [17,]   -3.559845    0.68430692   4.222375
##  [18,]   -3.896607    0.34754470   3.885613
##  [19,]   -3.345721    0.89843079   4.436499
##  [20,]   -3.571092    0.67305934   4.211127
##  [21,]   -3.504491    0.73966026   4.277728
```

In this *Wright Map* display, the results from the RS-MFRM analysis of the style ratings are summarized graphically. The figure is organized as follows:

- Units on the logit scale are shown on the far-right axis of the plot (labeled *Logits*).

- The left-most panel shows a histogram of student locations on the logit scale that represents the latent variable.

- The second panel from the left shows the distribution of subgroups on the logit scale. There are only two subgroups in our analysis.

The large central panel shows the rating scale category threshold estimates specific to each rater on the logit scale that represents the latent variable. Light gray diamond shapes show the logit scale location of the threshold estimates for each rater, as labeled on the x-axis. Thresholds are labeled using integers that show the threshold number. In our example, τ_1 is the threshold between rating scale categories $x = 0$ and $x = 1$, τ_2 is the threshold between rating scale categories $x = 1$ and $x = 2$, and τ_3 is the threshold between rating scale categories $x = 2$ and $x = 3$. Because we used a RS model formulation, the distance between adjacent thresholds is the same for all of the raters in the analysis.

Even though it is not appropriate to fully interpret item and person locations on the logit scale until there is evidence of acceptable model-data fit, we recommend examining the Wright Map during the preliminary stages of a MFRM analysis to get a general sense of the model results and to identify any potential scoring or data entry errors.

The Wright Map suggests that, on average, the students are located higher on the logit scale compared to the average rater threshold locations. In addition, there appears to be a relatively wide spread of student and rater locations on the logit scale, such that the style writing assessment appears to be a useful tool for identifying differences in students' writing achievement related to style as well as differences in rater severity. The subgroup locations are close together, suggesting that there is not a substantial difference in the logit-scale locations between students in either language subgroup.

Evaluate Model-Data Fit

Because the MFRM analyses result in notably different output structures from the other models in this book, we demonstrate a procedure for evaluating model-data fit specific to the MFRM with the TAM package. These procedures generally follow the methods that we presented in Chapter 3. We encourage readers to refer to Chapter 3 for more background information and details about model-data fit analysis procedures in the context of Rasch measurement theory.

Unidimensionality

First, we will evaluate the MFRM requirement for unidimensionality using the procedure that we demonstrated in Chapter 3. We will evaluate this requirement by examining the residuals for evidence of potential secondary latent variables in the rating data.

First, we will extract the model residuals using the IRT.residuals() function. We will save the results in an object called **resids**.

```
resids <- IRT.residuals(RS_MFR_model)
```

With the MFRM in *TAM*, residuals are presented separately for each level of the explanatory facets. We need to do some manipulation to construct a typical residual matrix.

```
# Extract the raw residuals from the residuals object
:
r <- as.data.frame(resids$residuals)

# Save the residuals in a matrix:
resid.matrix <- matrix(data = NA, nrow = nrow(style),
    ncol = ncol(ratings))

ngroups <- nrow(subgroup.estimates)

for(rater.number in 1:ncol(ratings)){
  group.raters <- NULL

  for(group in 1:ngroups){
    group.raters[group] <- paste("rater_", rater.
        number, "-", "language", group, sep = "")
  }

  rater <- subset(r, select = group.raters)

  resid.matrix[, rater.number] <- rowSums(rater, na.
      rm = TRUE)
}
```

The resulting residual matrix (**resid.matrix**) contains unstandardized residuals for each rater in combination with each student. We can request a summary of the residuals using the **summary()** function.

```
summary(resid.matrix)
```

```
##         V1                      V2
## Min.   :-0.8720729    Min.   :-1.4209183
## 1st Qu.:-0.2549334    1st Qu.:-0.2744779
## Median :-0.0076902    Median : 0.0132244
## Mean   : 0.0005486    Mean   : 0.0005993
```

```
##    3rd Qu.: 0.2620248    3rd Qu.: 0.2920722
##    Max.   : 0.8322741    Max.   : 1.1051632
##          V3                      V4
##    Min.   :-1.4791267    Min.   :-1.2272450
##    1st Qu.:-0.2851512    1st Qu.:-0.2748731
##    Median :-0.0509807    Median :-0.0406553
##    Mean   :-0.0001347    Mean   : 0.0000747
##    3rd Qu.: 0.2781122    3rd Qu.: 0.2935872
##    Max.   : 1.3321391    Max.   : 1.1978115
##          V5                      V6
##    Min.   :-1.652838     Min.   :-1.6572936
##    1st Qu.:-0.331711     1st Qu.:-0.2807091
##    Median : 0.030419     Median : 0.0109285
##    Mean   : 0.001325     Mean   : 0.0006332
##    3rd Qu.: 0.376558     3rd Qu.: 0.3427064
##    Max.   : 1.161870     Max.   : 1.3303463
##          V7                      V8
##    Min.   :-1.8408973    Min.   :-1.2572054
##    1st Qu.:-0.3348203    1st Qu.:-0.2642677
##    Median : 0.0032260    Median :-0.0403967
##    Mean   : 0.0004644    Mean   :-0.0002966
##    3rd Qu.: 0.3692606    3rd Qu.: 0.3012835
##    Max.   : 1.1287309    Max.   : 1.2138758
##          V9                      V10
##    Min.   :-1.490954     Min.   :-1.697046
##    1st Qu.:-0.260770     1st Qu.:-0.364370
##    Median :-0.002861     Median : 0.048410
##    Mean   :-0.001895     Mean   : 0.001479
##    3rd Qu.: 0.291447     3rd Qu.: 0.320855
##    Max.   : 1.196117     Max.   : 1.588882
##          V11                     V12
##    Min.   :-1.7755510    Min.   :-1.1677259
##    1st Qu.:-0.2788164    1st Qu.:-0.2651588
##    Median : 0.0054078    Median :-0.0076902
##    Mean   : 0.0004475    Mean   : 0.0005486
##    3rd Qu.: 0.2926797    3rd Qu.: 0.3079294
##    Max.   : 1.1316914    Max.   : 1.2168715
##          V13                     V14
##    Min.   :-1.2807091    Min.   :-1.3077252
##    1st Qu.:-0.2807091    1st Qu.:-0.2640100
##    Median :-0.0186308    Median :-0.0280155
##    Mean   : 0.0006332    Mean   :-0.0002784
##    3rd Qu.: 0.2832192    3rd Qu.: 0.3412485
##    Max.   : 2.1162376    Max.   : 1.2598161
##          V15                     V16
```

```
## Min.    :-1.514814    Min.    :-1.2111909
## 1st Qu.:-0.353619    1st Qu.:-0.2631766
## Median :  0.027757    Median :-0.0283414
## Mean    :  0.001402    Mean    :-0.0005784
## 3rd Qu.:  0.333995    3rd Qu.:  0.2917640
## Max.    :  1.659905    Max.    :  1.5083632
##        V17                    V18
## Min.    :-1.5034851    Min.    :-1.2532089
## 1st Qu.:-0.2427598    1st Qu.:-0.3478416
## Median :-0.0254666    Median :-0.0252699
## Mean    :-0.0005978    Mean    :-0.0000467
## 3rd Qu.:  0.2945528    3rd Qu.:  0.3573667
## Max.    :  0.9947392    Max.    :  1.7823467
##        V19                    V20
## Min.    :-1.9707405    Min.    :-1.3069233
## 1st Qu.:-0.2908683    1st Qu.:-0.2577943
## Median :-0.0163420    Median :-0.0403857
## Mean    :-0.0009824    Mean    :-0.0005784
## 3rd Qu.:  0.3191823    3rd Qu.:  0.2760813
## Max.    :  1.3759946    Max.    :  1.6930767
##        V21
## Min.    :-1.3787713
## 1st Qu.:-0.3494886
## Median :-0.0250489
## Mean    :-0.0009128
## 3rd Qu.:  0.3611649
## Max.    :  1.1498697
```

Next, we will calculate standardized residuals and save them in a matrix.

```
# Extract standardized residuals from the resids
    object:
s <- as.data.frame(resids$stand_residuals)

# Save the standardized residuals in a matrix:
std.resid.matrix <- matrix(data = NA, nrow = nrow(
    style), ncol = ncol(ratings))

ngroups <- nrow(subgroup.estimates)

for(rater.number in 1:ncol(ratings)){
  group.raters <- NULL

  for(group in 1:ngroups){
    group.raters[group] <- paste("rater_", rater.
      number, "-", "language", group, sep = "")
```

```
    }

    rater <- subset(s, select = group.raters)

    std.resid.matrix[, rater.number] <- rowSums(rater,
      na.rm = TRUE)
}
```

The resulting standardized residual matrix (std.resid.matrix) contains standardized residuals for each rater in combination with each student. We can request a summary of the standardized residuals using the summary() function.

```
summary(std.resid.matrix)
```

```
##          V1                      V2
##  Min.   :-1.905508     Min.    :-2.673927
##  1st Qu.:-0.506572     1st Qu.:-0.527221
##  Median :-0.017429     Median : 0.026156
##  Mean   : 0.001835     Mean    : 0.000882
##  3rd Qu.: 0.539578     3rd Qu.: 0.592587
##  Max.   : 1.614373     Max.    : 2.232782
##          V3                      V4
##  Min.   :-2.804533     Min.    :-2.351961
##  1st Qu.:-0.547302     1st Qu.:-0.545774
##  Median :-0.116122     Median :-0.091645
##  Mean   :-0.005136     Mean    :-0.002623
##  3rd Qu.: 0.600033     3rd Qu.: 0.561215
##  Max.   : 2.480104     Max.    : 2.294916
##          V5                      V6
##  Min.   :-3.550681     Min.    :-4.632938
##  1st Qu.:-0.643632     1st Qu.:-0.574105
##  Median : 0.154482     Median : 0.038452
##  Mean   :-0.000654     Mean    :-0.001968
##  3rd Qu.: 0.733945     3rd Qu.: 0.699961
##  Max.   : 2.445845     Max.    : 2.558106
##          V7                      V8
##  Min.   :-5.001028     Min.    :-2.395289
##  1st Qu.:-0.653402     1st Qu.:-0.553848
##  Median : 0.007315     Median :-0.091069
##  Mean   : 0.002132     Mean    :-0.004299
##  3rd Qu.: 0.723834     3rd Qu.: 0.630877
##  Max.   : 2.621032     Max.    : 2.342300
##          V9                      V10
##  Min.   :-2.76309      Min.    :-3.631079
##  1st Qu.:-0.53408      1st Qu.:-0.714225
##  Median :-0.03185      Median : 0.107363
```

```
##    Mean     :-0.01786    Mean    : 0.007239
##    3rd Qu.:  0.55894     3rd Qu.:  0.648215
##    Max.     : 2.50557    Max.    : 2.991603
##          V11                      V12
##    Min.     :-3.379879   Min.     :-2.4268836
##    1st Qu.:-0.559049     1st Qu.:-0.5582567
##    Median :  0.012261    Median :-0.0174290
##    Mean     : 0.000516   Mean     :-0.0009097
##    3rd Qu.:  0.600774    3rd Qu.:  0.6148613
##    Max.     : 2.279635   Max.     : 2.5338432
##          V13                      V14
##    Min.     :-2.43750    Min.     :-2.47939
##    1st Qu.:-0.57410      1st Qu.:-0.52845
##    Median :-0.04218      Median :-0.09599
##    Mean     : 0.01468    Mean     :-0.00148
##    3rd Qu.:  0.58357     3rd Qu.:  0.66432
##    Max.     : 4.65120    Max.     : 3.34157
##          V15                      V16
##    Min.     :-3.633244   Min.     :-2.32838
##    1st Qu.:-0.690213     1st Qu.:-0.52702
##    Median :  0.128835    Median :-0.08551
##    Mean     :-0.009865   Mean     : 0.01256
##    3rd Qu.:  0.631985    3rd Qu.:  0.58627
##    Max.     : 3.202808   Max.     : 2.99223
##          V17                      V18
##    Min.     :-2.866723   Min.     :-2.47200
##    1st Qu.:-0.485482     1st Qu.:-0.65646
##    Median :-0.080362     Median :-0.05715
##    Mean     : 0.006671   Mean     : 0.01111
##    3rd Qu.:  0.586370    3rd Qu.:  0.70809
##    Max.     : 4.899036   Max.     : 4.29775
##          V19                      V20
##    Min.     :-3.894471   Min.     :-3.13528
##    1st Qu.:-0.584566     1st Qu.:-0.52702
##    Median :-0.032333     Median :-0.11528
##    Mean     :-0.003165   Mean     :-0.02053
##    3rd Qu.:  0.622024    3rd Qu.:  0.58068
##    Max.     : 4.095676   Max.     : 3.31535
##          V21
##    Min.     :-2.621914
##    1st Qu.:-0.720283
##    Median :-0.086063
##    Mean     :-0.003733
##    3rd Qu.:  0.709318
##    Max.     : 2.637999
```

Next, we will calculate the variance in observations due to Rasch-model-estimated locations.

```
# Variance of the observations: VO
observations.vector <- as.vector(as.matrix(ratings))
VO <- var(observations.vector)
```

```
# Variance of the residuals: VR
residuals.vector <- as.vector(resid.matrix)
VR <- var(residuals.vector)
```

```
# Raw variance explained by Rasch measures: (VO - VR)
    /VO
(VO - VR)/VO
```

```
## [1] 0.7418448
```

```
# Express the result as a percent:
((VO - VR)/VO) * 100
```

```
## [1] 74.18448
```

Our analysis indicates that approximately 74.18% of the variance in ratings can be explained by the MFRM estimates of student, subgroup, and rater locations on the logit scale that represents the latent variable.

Principal Components Analysis of Standardized Residual Correlations

Next, we will evaluate the MFRM requirement for unidimensionality using a principal components analysis (PCA) of standardized residual correlations.

```
pca <- pca(std.resid.matrix, nfactors = ncol(ratings)
    , rotate = "none")
```

```
contrasts <- c(pca$values[1], pca$values[2], pca$
    values[3], pca$values[4], pca$values[5])
```

```
plot(contrasts, ylab = "Eigenvalues for Contrasts",
    xlab = "Contrast Number", main = "Contrasts from
    PCA Standardized Residual Correlations", ylim = c
    (0, 2))
```

Contrasts from PCA of Standardized Residual Correlations

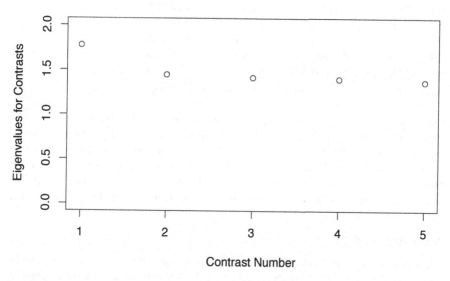

In this example, all of the contrasts have eigenvalues that are smaller than Linacre's (2016)'s critical value of 2.00. This result suggests that the correlations among the model residuals primarily reflect randomness (i.e., noise)—thus providing evidence that the responses adequately adhere to the Rasch model requirement of unidimensionality.

Summaries of Residuals: Infit & Outfit Statistics

Next, we will evaluate model-data fit for individual elements of our facets (students, subgroups, and raters) using numeric summaries of the residuals associated with each element, as we have done in previous chapters.

Student fit

First, we will examine student fit using numeric infit and outfit statistics. We can request these statistics for each student using the `tam.personfit()` function. We will store the student fit results in an object called `student.fit`, and then request a summary of the results.

```
student.fit <- tam.personfit(RS_MFR_model)
summary(student.fit)
```

```
##    outfitPerson        outfitPerson_t
## Min.    :0.02316    Min.    :-4.4319
## 1st Qu.:0.68599    1st Qu.:-0.9745
## Median :0.86657    Median :-0.3590
## Mean    :0.89235    Mean    :-0.3164
## 3rd Qu.:1.05760    3rd Qu.: 0.2927
## Max.    :2.27289    Max.    : 3.5578
```

```
##    infitPerson          infitPerson_t
##   Min.   :0.02864    Min.    :-4.4127
##   1st Qu.:0.71529    1st Qu.:-0.9820
##   Median :0.87397    Median :-0.3395
##   Mean   :0.90160    Mean    :-0.3018
##   3rd Qu.:1.03086    3rd Qu.: 0.2107
##   Max.   :2.39705    Max.    : 3.6511
```

On average, the outfit and infit MSE statistics are slightly lower than the expected value of 1 (M outfit = 0.89, M infit = 0.90). The average values of the standardized fit statistics are also slightly lower than their expected value of 0 (M standardized outfit = -0.33, M standardized infit = -0.31). For both the standardized and unstandardized fit statistics, there is notable variability across the student sample. This result suggests that model-data fit varies for individual students.

Subgroup and Rater Fit

We can also examine model-data fit related to the subgroup and rater facets. In the *TAM* package, fit analysis for facets besides the object of measurement uses combinations of elements within facets. In our example, fit statistics are calculated for rater*subgroup combinations.

```
rater.subgroup.fit <- msq.itemfit(RS_MFR_model)
summary(rater.subgroup.fit)
```

```
## -----------------------------------------------------
## TAM 3.7-16 (2021-06-24 14:31:37)
## R version 4.1.0 (2021-05-18) x86_64, darwin17.0 |
##   nodename=XXXX | login=root
##
## Date of Analysis: 2021-10-27 13:46:52
## Time difference of 0.04882312 secs
## Computation time: 0.04882312
##
## MSQ item fit statitics (Function 'msq.itemfit')
##
## Call:
## msq.itemfit(object = RS_MFR_model)
##
## ****************************************************
##
## Summary outfit and infit statistic
##        fit      M      SD
## 1 Outfit 0.948 0.187
## 2  Infit 0.956 0.175
##
```

```
## ************************************************************
##
## Outfit and infit statistic
##                      item fitgroup  Outfit  Outfit_t
## 1    rater_1-language1         1    0.538    -4.842
## 2    rater_1-language2         2    0.515    -5.661
## 3    rater_10-language1        3    0.994    -0.004
## 4    rater_10-language2        4    1.163     1.417
## 5    rater_11-language1        5    0.707    -2.821
## 6    rater_11-language2        6    0.968    -0.271
## 7    rater_12-language1        7    0.809    -1.747
## 8    rater_12-language2        8    0.859    -1.387
## 9    rater_13-language1        9    0.633    -3.645
## 10   rater_13-language2       10    0.989    -0.067
## 11   rater_14-language1       11    0.945    -0.447
## 12   rater_14-language2       12    1.062     0.614
## 13   rater_15-language1       13    1.095     0.794
## 14   rater_15-language2       14    1.161     1.413
## 15   rater_16-language1       15    0.749    -2.340
## 16   rater_16-language2       16    0.999     0.031
## 17   rater_17-language1       17    0.683    -3.041
## 18   rater_17-language2       18    0.889    -1.035
## 19   rater_18-language1       19    1.092     0.821
## 20   rater_18-language2       20    1.240     2.167
## 21   rater_19-language1       21    0.996     0.007
## 22   rater_19-language2       22    1.096     0.880
## 23    rater_2-language1       23    0.769    -2.143
## 24    rater_2-language2       24    0.840    -1.586
## 25   rater_20-language1       25    0.871    -1.125
## 26   rater_20-language2       26    0.963    -0.312
## 27   rater_21-language1       27    1.195     1.615
## 28   rater_21-language2       28    1.240     2.101
## 29    rater_3-language1       29    0.840    -1.443
## 30    rater_3-language2       30    1.002     0.051
## 31    rater_4-language1       31    0.772    -2.143
## 32    rater_4-language2       32    0.840    -1.592
## 33    rater_5-language1       33    1.140     1.140
## 34    rater_5-language2       34    1.187     1.635
## 35    rater_6-language1       35    1.047     0.437
## 36    rater_6-language2       36    1.021     0.238
## 37    rater_7-language1       37    1.236     1.955
## 38    rater_7-language2       38    1.090     0.871
## 39    rater_8-language1       39    1.008     0.111
## 40    rater_8-language2       40    0.819    -1.793
## 41    rater_9-language1       41    0.970    -0.197
```

```
## 42   rater_9-language2          42  0.790    -1.861
##      Outfit_p Infit Infit_t Infit_p
## 1      0.000 0.569  -4.906   0.000
## 2      0.000 0.550  -5.551   0.000
## 3      0.997 1.010   0.136   0.892
## 4      0.156 1.189   1.870   0.062
## 5      0.005 0.745  -2.643   0.008
## 6      0.786 1.000   0.029   0.977
## 7      0.081 0.835  -1.647   0.100
## 8      0.165 0.860  -1.478   0.139
## 9      0.000 0.660  -3.696   0.000
## 10     0.946 0.997  -0.004   0.997
## 11     0.655 0.960  -0.341   0.733
## 12     0.539 1.058   0.610   0.542
## 13     0.427 1.070   0.701   0.483
## 14     0.158 1.073   0.772   0.440
## 15     0.019 0.779  -2.191   0.028
## 16     0.975 0.983  -0.136   0.892
## 17     0.002 0.689  -3.218   0.001
## 18     0.300 0.874  -1.302   0.193
## 19     0.412 1.100   0.949   0.343
## 20     0.030 1.203   1.977   0.048
## 21     0.995 1.009   0.123   0.902
## 22     0.379 1.133   1.318   0.188
## 23     0.032 0.784  -2.212   0.027
## 24     0.113 0.833  -1.794   0.073
## 25     0.261 0.883  -1.100   0.271
## 26     0.755 0.953  -0.456   0.648
## 27     0.106 1.188   1.680   0.093
## 28     0.036 1.268   2.535   0.011
## 29     0.149 0.851  -1.443   0.149
## 30     0.959 0.982  -0.148   0.882
## 31     0.032 0.781  -2.205   0.027
## 32     0.111 0.836  -1.749   0.080
## 33     0.254 1.155   1.476   0.140
## 34     0.102 1.165   1.653   0.098
## 35     0.662 1.047   0.480   0.631
## 36     0.812 1.049   0.534   0.593
## 37     0.051 1.148   1.389   0.165
## 38     0.384 1.116   1.182   0.237
## 39     0.911 1.039   0.392   0.695
## 40     0.073 0.852  -1.559   0.119
## 41     0.844 1.031   0.322   0.748
## 42     0.063 0.853  -1.503   0.133
```

As needed, researchers can also examine model-data fit statistics specific to subgroups of examinees. This can be accomplished by calculating summary statistics of person fit statistics within examinee subgroups. The code below merges the person fit results with student subgroup identification numbers, and then calculates summary statistics and produces boxplots of the fit statistics by group.

```
fit_with_subgroups <- cbind.data.frame(style$language
  , student.fit)

fit_group_1 <- subset(fit_with_subgroups, fit_with_
  subgroups$`style$language` == 1)

summary(fit_group_1)
```

```
## style$language   outfitPerson
## Min.    :1       Min.    :0.02316
## 1st Qu.:1        1st Qu.:0.68466
## Median :1        Median :0.82739
## Mean    :1       Mean    :0.85283
## 3rd Qu.:1        3rd Qu.:1.02341
## Max.    :1       Max.    :2.07327
## outfitPerson_t       infitPerson
## Min.    :-4.4319     Min.    :0.02864
## 1st Qu.:-0.9418      1st Qu.:0.70913
## Median :-0.4485      Median :0.83743
## Mean    :-0.4074     Mean    :0.85791
## 3rd Qu.: 0.1809      3rd Qu.:0.99769
## Max.    : 3.3507     Max.    :2.02119
## infitPerson_t
## Min.    :-4.41269
## 1st Qu.:-0.97773
## Median :-0.44301
## Mean    :-0.40876
## 3rd Qu.: 0.08487
## Max.    : 3.28558
```

```
fit_group_2 <- subset(fit_with_subgroups, fit_with_
  subgroups$`style$language` == 2)

summary(fit_group_2)
```

```
## style$language   outfitPerson
## Min.    :2       Min.    :0.1389
## 1st Qu.:2        1st Qu.:0.6915
## Median :2        Median :0.8848
```

```
##   Mean     :2        Mean      :0.9271
##   3rd Qu.:2           3rd Qu.:1.0873
##   Max.     :2         Max.      :2.2729
##   outfitPerson_t         infitPerson
##   Min.     :-3.5105    Min.     :0.1599
##   1st Qu.:-0.9962      1st Qu.:0.7330
##   Median :-0.3268      Median :0.8932
##   Mean     :-0.2365    Mean      :0.9400
##   3rd Qu.: 0.4161      3rd Qu.:1.0843
##   Max.     : 3.5578    Max.      :2.3971
##   infitPerson_t
##   Min.     :-3.4856
##   1st Qu.:-0.9809
##   Median :-0.2919
##   Mean     :-0.2078
##   3rd Qu.: 0.4148
##   Max.     : 3.6511
```

```
# Boxplots for MSE fit statistics:
boxplot(fit_group_1$outfitPerson, fit_group_2$
   outfitPerson,
        fit_group_1$infitPerson, fit_group_2$
           infitPerson,
        names = c("Group \nOutfit MSE", "Group \
           nOutfit MSE",
                  "Group \nInfit MSE", "Group \nInfit
                     MSE"),
        col = c("grey", "white", "grey", "white"),
        main = "MSE Fit Statistics for English-not-
           Best-Language (Group) \nand English-Best-
           Language (Group) Students",
        cex.main = .8,
        ylab = "MSE Fit Statistic", xlab = "Student
           Subgroup")
```

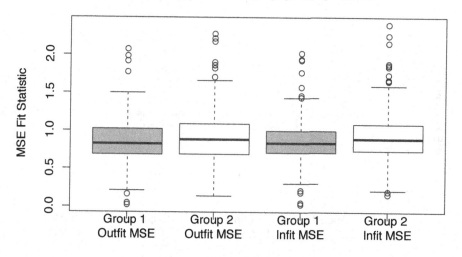

MSE Fit Statistics for English–not–Best–Language (Group 1) and English–Best–Language (Group 2) Students

```
# Boxplots for standardized fit statistics:
boxplot(fit_group_1$outfitPerson_t, fit_group_2$
    outfitPerson_t,
        fit_group_1$infitPerson_t, fit_group_2$
            infitPerson_t,
        names = c("Group \nStd. Outfit", "Group \nStd
            . Outfit",
                "Group \nStd. Infit", "Group \nStd.
                    Infit"),
        col = c("grey", "white", "grey", "white"),
        main = "Standardized Fit Statistics for
            English-not-Best-Language (Group) \nand
            English-Best-Language (Group) Students",
        cex.main = .8,
        ylab = "MSE Fit Statistic", xlab = "Student
            Subgroup")
```

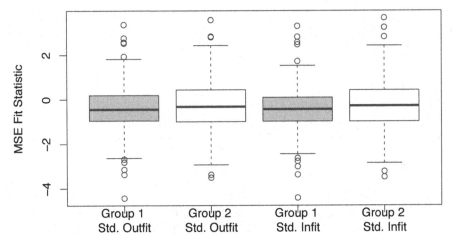

Finally, it may be useful to examine fit statistics as they apply to individual raters. This can be accomplished by extracting rater-specific fit statistics within each subgroup. The code below calculates rater fit statistics within each subgroup.

```
ngroups <- nrow(subgroup.estimates)

rater.fit <- matrix(data = NA, nrow = ncol(ratings),
    ncol = (ngroups * 4) + 1 )

for(rater.number in 1:ncol(ratings)){

    # calculate rater-specific fit statistics:
    rater.outfit <- rater.subgroup.fit$itemfit$Outfit
        [((rater.number*ngroups) - (ngroups - 1)) : (
        rater.number*ngroups)]

        rater.infit <- rater.subgroup.fit$itemfit$Infit
            [((rater.number*ngroups) - (ngroups - 1)) : (
            rater.number*ngroups)]

        rater.std.outfit <- rater.subgroup.fit$itemfit$
            Outfit_t[((rater.number*ngroups) - (ngroups -
            1)) : (rater.number*ngroups)]
```

```
rater.std.infit <- rater.subgroup.fit$itemfit$Infit
    _t[((rater.number*ngroups) - (ngroups - 1)) : (
    rater.number*ngroups)]

    # add the fit statistics to the matrix:
    rater.fit[rater.number, ] <-  c(rater.number, rater
        .outfit, rater.infit,
                                    rater.std.outfit,
                                        rater.std.infit)
}


# Convert the rater fit results to a dataframe object
    and add meaningful column names:

rater.fit_results <- as.data.frame(rater.fit)

infit_mse_labels <- NULL
for(group in 1:ngroups){
    infit_mse_labels[group] <- paste("Infit_MSE_Group",
        group, sep = "")
}

outfit_mse_labels <- NULL
for(group in 1:ngroups){
    outfit_mse_labels[group] <- paste("Outfit_MSE_Group
        ", group, sep = "")
    }

std_infit_mse_labels <- NULL
for(group in 1:ngroups){
    std_infit_mse_labels[group] <- paste("Std.Infit_MSE
        _Group", group, sep = "")
}

std_outfit_mse_labels <- NULL
for(group in 1:ngroups){
    std_outfit_mse_labels[group] <- paste("Std.Outfit_
        MSE_Group", group, sep = "")
    }

names(rater.fit_results) <- c("Rater", outfit_mse_
    labels, infit_mse_labels,
                            std_outfit_mse_labels,
```

```
                                       std_infit_mse_labels
                                       )
```

```
# Display the rater fit results for the first six
  raters using the head() function:
head(rater.fit_results)
```

```
##    Rater Outfit_MSE_Group1 Outfit_MSE_Group2
## 1    1         0.5378319         0.5152339
## 2    2         0.9943884         1.1627861
## 3    3         0.7072920         0.9684632
## 4    4         0.8087563         0.8586822
## 5    5         0.6328367         0.9893404
## 6    6         0.9454056         1.0622633
##    Infit_MSE_Group1 Infit_MSE_Group2
## 1        0.5685180         0.5500132
## 2        1.0104348         1.1889506
## 3        0.7454948         0.9996399
## 4        0.8348633         0.8604496
## 5        0.6602235         0.9965140
## 6        0.9602931         1.0577594
##    Std.Outfit_MSE_Group1 Std.Outfit_MSE_Group2
## 1           -4.84231202           -5.66063470
## 2           -0.00409796            1.41721943
## 3           -2.82054661           -0.27109066
## 4           -1.74675007           -1.38725250
## 5           -3.64467726           -0.06720607
## 6           -0.44670525            0.61361924
##    Std.Infit_MSE_Group1 Std.Infit_MSE_Group2
## 1          -4.9062354         -5.551375656
## 2           0.1355469          1.869833670
## 3          -2.6428614          0.028685263
## 4          -1.6466281         -1.477855312
## 5          -3.6957750         -0.003650316
## 6          -0.3414119          0.610061859
```

Now that we have created a data frame with the rater-specific fit statistics, we can summarize the results.

```
summary(rater.fit_results)
```

```
##       Rater     Outfit_MSE_Group1 Outfit_MSE_Group2
##   Min.   : 1    Min.   :0.5378    Min.   :0.5152
##   1st Qu.: 6    1st Qu.:0.7694    1st Qu.:0.8587
##   Median :11    Median :0.9454    Median :0.9995
##   Mean   :11    Mean   :0.9091    Mean   :0.9873
```

```
## 3rd Qu.:16      3rd Qu.:1.0466      3rd Qu.:1.0958
## Max.     :21    Max.     :1.2359    Max.     :1.2400
## Infit_MSE_Group1  Infit_MSE_Group2
## Min.     :0.5685  Min.     :0.5500
## 1st Qu.:0.7814    1st Qu.:0.8604
## Median :0.9603    Median :0.9965
## Mean     :0.9207  Mean     :0.9918
## 3rd Qu.:1.0470    3rd Qu.:1.1159
## Max.     :1.1878  Max.     :1.2682
## Std.Outfit_MSE_Group1  Std.Outfit_MSE_Group2
## Min.     :-4.8423      Min.     :-5.66064
## 1st Qu.:-2.1434        1st Qu.:-1.38725
## Median :-0.4467        Median : 0.03104
## Mean     :-0.9075      Mean     :-0.19763
## 3rd Qu.: 0.4368        3rd Qu.: 0.87994
## Max.     : 1.9554      Max.     : 2.16736
## Std.Infit_MSE_Group1   Std.Infit_MSE_Group2
## Min.     :-4.9062      Min.     :-5.55138
## 1st Qu.:-2.2053        1st Qu.:-1.47785
## Median :-0.3414        Median :-0.00365
## Mean     :-0.8551      Mean     :-0.15246
## 3rd Qu.: 0.4802        3rd Qu.: 1.18162
## Max.     : 1.6800      Max.     : 2.53485
```

Graphical Displays of Residuals

Continuing our fit analysis, we will construct plots of standardized residuals associated with individual raters. These plots can highlight patterns in unexpected and expected responses that can be useful for understanding responses and interpreting results specific to individual raters. In other applications of the MFRM, researchers can construct similar plots for other facets.

Earlier in this chapter, we stored the standardized residuals in an object called std.resid.matrix. We will use this object to create plots for individual raters via a for-loop. For brevity, we have only included plots for the first three raters in this book. The specific raters to be plotted can be controlled by changing the items included in the raters.to.plot object.

```
# Before constructing the plots, find the maximum and
    minimum values of the standardized residuals to
    set limits for the axes:
max.resid <- ceiling(max(std.resid.matrix))
min.resid <- ceiling(min(std.resid.matrix))

# The code below will produce plots of standardized
    residuals for selected raters as listed in raters.
    to.plot:
```

```
raters.to.plot <- c(1:3)

for(rater.number in raters.to.plot){
  plot(std.resid.matrix[, rater.number], ylim = c(min
    .resid, max.resid),
      main = paste("Standardized Residuals for Rater
        ", rater.number, sep = ""),
      ylab = "Standardized Residual", xlab = "Person
        Index")
  abline(h = 0, col = "blue")
  abline(h=2, lty = 2, col = "red")
  abline(h=-2, lty = 2, col = "red")

  legend("topright", c("Std. Residual", "Observed =
    Expected", "+/- SD"), pch = c(1, NA, NA),
      lty = c(NA, 1, 2),
      col = c("black", "blue", "red"), cex = .8)
}
```

Standardized Residuals for Rater 1

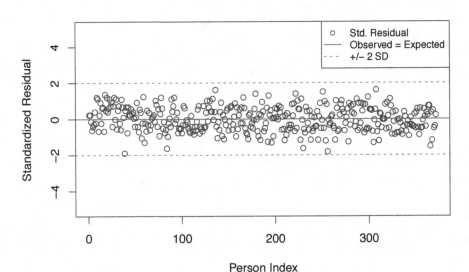

Standardized Residuals for Rater 2

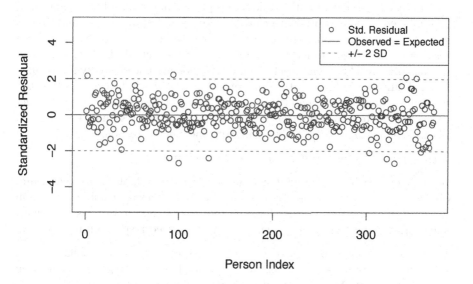

Standardized Residuals for Rater 3

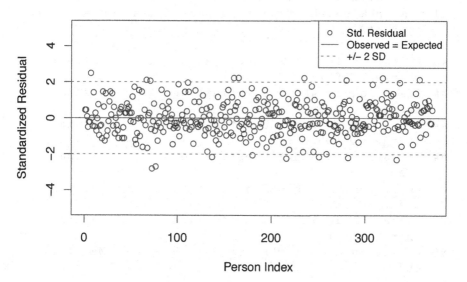

A separate plot is produced for each rater. In each plot, the y-axis shows values of the standardized residuals, and the x-axis shows the students, ordered by their relative position in the data set. Open-circle plotting symbols show the standardized residual associated with each student's rating from the rater of interest.

Horizontal lines are used to assist in the interpretation of the values of the standardized residuals. First, a solid line is plotted at a value of 0; standardized residuals equal to zero indicate that the observed response was equal to the model-expected response given student and rater locations. Standardized residuals that are greater than zero indicate unexpectedly high ratings, and standardized residuals that are less than zero indicate unexpectedly low ratings. Dashed lines are plotted at values of +2 and -2 to indicate standardized residuals that are two standard deviations above or below model expectations, respectively. Researchers often interpret standardized residuals that exceed +/- 2 as indicating statistically significant unexpected responses.

Expected and Observed Response Functions

As a final step in our fit analysis, we will construct plots of expected and observed response functions. By default, the *TAM* package combines the item facet (in this case, raters), with levels of the other facets (in this case, language subgroups) when constructing expected and observed response function plots.

For brevity, we only plot the expected and observed response functions for three selected rater*item combinations. Readers can adjust the `items` = specification to construct plots for elements of interest for their analyses.

```
raters.to.plot <- c(1:3)
plot(RS_MFR_model, type = "expected", items = raters.
    to.plot)
```

```
## Iteration in WLE/MLE estimation  1   | Maximal
    change  2.9877
## Iteration in WLE/MLE estimation  2   | Maximal
    change  2.1965
## Iteration in WLE/MLE estimation  3   | Maximal
    change  0.605
## Iteration in WLE/MLE estimation  4   | Maximal
    change  0.1983
## Iteration in WLE/MLE estimation  5   | Maximal
    change  0.009
## Iteration in WLE/MLE estimation  6   | Maximal
    change  5e-04
## Iteration in WLE/MLE estimation  7   | Maximal
    change  0
## ----
##  WLE Reliability= 0.977
```

Expected Scores Curve – Item rater_1–language1

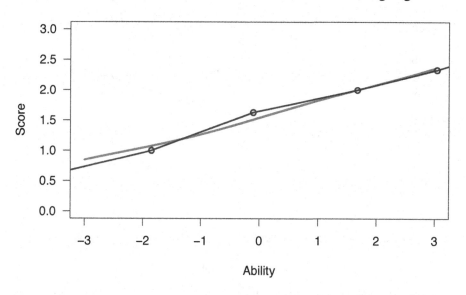

Expected Scores Curve – Item rater_1–language2

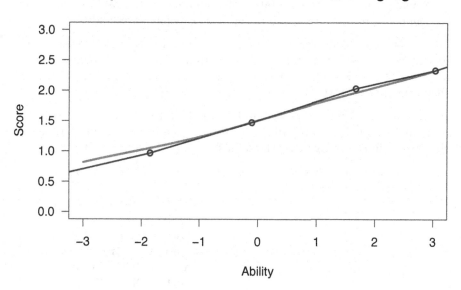

Expected Scores Curve – Item rater_10–language1

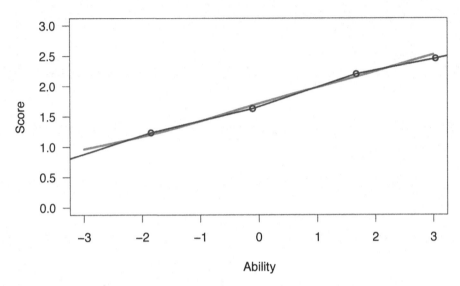

Summarize the Results in Tables

As a final step, we will create tables that summarize the calibrations of the students, subgroups, raters, and rating scale category thresholds.

Table 1 provides an overview of the logit scale locations, standard errors and fit statistics for all of the facets in the analysis. This table provides a quick overview of the location estimates and numeric model-data fit statistics for the facets in a MFRM.

Because of the estimation procedure for the MFRM in *TAM*, fit statistics are combined for the item facet and other facets. As a result, the fit statistics in this table will be the same for the rater facet and the subgroup facets.

```
RS_MFRM_summary.table.statistics <- c("Logit Scale
    Location Mean",
                        "Logit Scale Location
                            SD",
                        "Standard Error Mean",
                        "Standard Error SD",
                        "Outfit MSE Mean",
                        "Outfit MSE SD",
                        "Infit MSE Mean",
                        "Infit MSE SD",
                        "Std. Outfit Mean",
                        "Std. Outfit SD",
                        "Std. Infit Mean",
                        "Std. Infit SD")
```

```
RS_MFRM_student.summary.results <- rbind(mean(student
    .locations_RSMFR$theta),
                        sd(student.locations_
                            RSMFR$theta),
                        mean(student.locations_
                            RSMFR$se),
                        sd(student.locations_
                            RSMFR$se),
                        mean(student.fit$
                            outfitPerson),
                        sd(student.fit$
                            outfitPerson),
                        mean(student.fit$
                            infitPerson),
                        sd(student.fit$
                            infitPerson),
                        mean(student.fit$
                            outfitPerson_t),
                        sd(student.fit$
                            outfitPerson_t),
                        mean(student.fit$
                            infitPerson_t),
                        sd(student.fit$
                            infitPerson_t))

RS_MFRM_subgroup.summary.results <- rbind(mean(
    subgroup.estimates$xsi),
                        sd(subgroup.estimates$
                            xsi),
                        mean(subgroup.estimates
                            $se.xsi),
                        sd(subgroup.estimates$
                            se.xsi),
                        mean(rater.subgroup.fit
                            $itemfit$Outfit),
                        sd(rater.subgroup.fit$
                            itemfit$Outfit),
                        mean(rater.subgroup.fit
                            $itemfit$Infit),
                        sd(rater.subgroup.fit$
                            itemfit$Infit),
                        mean(rater.subgroup.fit
                            $itemfit$Outfit_t),
```

```
                                        sd(rater.subgroup.fit$
                                            itemfit$Outfit_t),
                                        mean(rater.subgroup.fit
                                            $itemfit$Infit_t),
                                        sd(rater.subgroup.fit$
                                            itemfit$Infit_t))

RS_MFRM_rater.summary.results <- rbind(mean(rater.
    estimates$xsi),
                                    sd(rater.estimates$xsi),
                                    mean(rater.estimates$se
                                        .xsi),
                                    sd(rater.estimates$se.
                                        xsi),
                                    mean(rater.subgroup.fit
                                        $itemfit$Outfit),
                                    sd(rater.subgroup.fit$
                                        itemfit$Outfit),
                                    mean(rater.subgroup.fit
                                        $itemfit$Infit),
                                    sd(rater.subgroup.fit$
                                        itemfit$Infit),
                                    mean(rater.subgroup.fit
                                        $itemfit$Outfit_t),
                                    sd(rater.subgroup.fit$
                                        itemfit$Outfit_t),
                                    mean(rater.subgroup.fit
                                        $itemfit$Infit_t),
                                    sd(rater.subgroup.fit$
                                        itemfit$Infit_t))


# Round the values for presentation in a table:
RS_MFRM_student.summary.results_rounded <- round(RS_
    MFRM_student.summary.results, digits = 2)

RS_MFRM_subgroup.summary.results_rounded <- round(RS_
    MFRM_subgroup.summary.results, digits = 2)

RS_MFRM_rater.summary.results_rounded <- round(RS_
    MFRM_rater.summary.results, digits = 2)


RS_MFRM_Table1 <- cbind.data.frame(RS_MFRM_summary.
    table.statistics,
```

```
                                        RS_MFRM_student.summary.
                                           results_rounded,
                                        RS_MFRM_subgroup.summary.
                                           results_rounded,
                                        RS_MFRM_rater.summary.
                                           results_rounded)


# add descriptive column labels:
names(RS_MFRM_Table1) <- c("Statistic", "Students", "
    Subgroups", "Raters")

# Print the table to the console
print.data.frame(RS_MFRM_Table1, row.names = FALSE)

##                         Statistic Students Subgroups
##      Logit Scale Location Mean        0.47      0.00
##       Logit Scale Location SD         3.17      0.09
##           Standard Error Mean         0.47      0.02
##            Standard Error SD          0.12      0.00
##             Outfit MSE Mean           0.89      0.95
##              Outfit MSE SD            0.35      0.19
##              Infit MSE Mean           0.90      0.96
##               Infit MSE SD            0.35      0.17
##          Std. Outfit Mean           -0.32     -0.55
##           Std. Outfit SD              1.13      1.82
##           Std. Infit Mean           -0.30     -0.50
##            Std. Infit SD              1.12      1.85
##   Raters
##      0.00
##      0.52
##      0.09
##      0.06
##      0.95
##      0.19
##      0.96
##      0.17
##     -0.55
##      1.82
##     -0.50
##      1.85
```

Table 2 summarizes the overall calibrations of individual raters. For data sets
with manageable sample sizes such as the style ratings example in this chapter,

we recommend reporting details about each element of explanatory facets (e.g., individual raters) in a table similar to this one.

```
# Calculate the average rating for each rater:
Avg_Rating <- apply(ratings, 2, mean)

# Combine rater calibration results in a table:

RS_MFRM_Table2 <- cbind.data.frame(c(1:ncol(ratings))
    ,
                            Avg_Rating,
                            rater.estimates$xsi,
                            rater_thresholds,
                            rater.fit_results[, -1])

names(RS_MFRM_Table2) <- c("Rater ID", "Average
    Rating", "Rater Location","Threshold 1", "
    Threshold 2", "Threshold 3", names(rater.fit_
    results[, -1]))

# Sort Table 2 by rater severity:
RS_MFRM_Table2 <- RS_MFRM_Table2[order(-RS_MFRM_
    Table2$`Rater Location`),]

# Round the numeric values (all columns except the
    first one) to 2 digits:
RS_MFRM_Table2[, -1] <- round(RS_MFRM_Table2[,-1],
    digits = 2)

# Print the table to the console
print.data.frame(RS_MFRM_Table2, row.names = FALSE)
```

```
##   Rater ID Average Rating Rater Location
##          9           1.35           1.12
##         19           1.45           0.66
##         21           1.49           0.50
##         17           1.51           0.45
##         16           1.51           0.44
##         20           1.51           0.44
##          8           1.55           0.27
##         14           1.55           0.26
##          3           1.57           0.17
##         18           1.59           0.11
##          4           1.60           0.03
##         11           1.66          -0.21
##          7           1.67          -0.22
```

```
##              1            1.68               -0.28
##             12            1.68               -0.28
##              2            1.69               -0.31
##              6            1.69               -0.34
##             13            1.69               -0.34
##              5            1.80               -0.78
##             15            1.81               -0.82
##             10            1.82               -0.87
##      Threshold 1  Threshold 2  Threshold 3
##            -2.89         1.35          4.89
##            -3.35         0.90          4.44
##            -3.50         0.74          4.28
##            -3.56         0.68          4.22
##            -3.57         0.67          4.21
##            -3.57         0.67          4.21
##            -3.74         0.50          4.04
##            -3.75         0.49          4.03
##            -3.84         0.40          3.94
##            -3.90         0.35          3.89
##            -3.98         0.27          3.81
##            -4.22         0.02          3.56
##            -4.23         0.01          3.55
##            -4.29        -0.04          3.49
##            -4.29        -0.04          3.49
##            -4.32        -0.08          3.46
##            -4.35        -0.10          3.44
##            -4.35        -0.10          3.44
##            -4.79        -0.54          3.00
##            -4.83        -0.59          2.95
##            -4.88        -0.63          2.91
##      Outfit_MSE_Group1  Outfit_MSE_Group2
##                   0.68               0.89
##                   1.24               1.09
##                   0.97               0.79
##                   1.14               1.19
##                   0.77               0.84
##                   1.01               0.82
##                   0.75               1.00
##                   1.20               1.24
##                   0.71               0.97
##                   1.05               1.02
##                   0.81               0.86
##                   1.00               1.10
##                   1.10               1.16
##                   0.54               0.52
```

```
##                   0.77                      0.84
##                   0.99                      1.16
##                   0.95                      1.06
##                   0.87                      0.96
##                   0.63                      0.99
##                   0.84                      1.00
##                   1.09                      1.24
##     Infit_MSE_Group1 Infit_MSE_Group2
##                   0.69                      0.87
##                   1.15                      1.12
##                   1.03                      0.85
##                   1.16                      1.17
##                   0.78                      0.84
##                   1.04                      0.85
##                   0.78                      0.98
##                   1.19                      1.27
##                   0.75                      1.00
##                   1.05                      1.05
##                   0.83                      0.86
##                   1.01                      1.13
##                   1.07                      1.07
##                   0.57                      0.55
##                   0.78                      0.83
##                   1.01                      1.19
##                   0.96                      1.06
##                   0.88                      0.95
##                   0.66                      1.00
##                   0.85                      0.98
##                   1.10                      1.20
##     Std.Outfit_MSE_Group1 Std.Outfit_MSE_Group2
##                     -3.04                     -1.04
##                      1.96                      0.87
##                     -0.20                     -1.86
##                      1.14                      1.63
##                     -2.14                     -1.59
##                      0.11                     -1.79
##                     -2.34                      0.03
##                      1.62                      2.10
##                     -2.82                     -0.27
##                      0.44                      0.24
##                     -1.75                     -1.39
##                      0.01                      0.88
##                      0.79                      1.41
##                     -4.84                     -5.66
##                     -2.14                     -1.59
```

```
##                          0.00                          1.42
##                         -0.45                          0.61
##                         -1.13                         -0.31
##                         -3.64                         -0.07
##                         -1.44                          0.05
##                          0.82                          2.17
##    Std.Infit_MSE_Group1 Std.Infit_MSE_Group2
##                         -3.22                         -1.30
##                          1.39                          1.18
##                          0.32                         -1.50
##                          1.48                          1.65
##                         -2.21                         -1.75
##                          0.39                         -1.56
##                         -2.19                         -0.14
##                          1.68                          2.53
##                         -2.64                          0.03
##                          0.48                          0.53
##                         -1.65                         -1.48
##                          0.12                          1.32
##                          0.70                          0.77
##                         -4.91                         -5.55
##                         -2.21                         -1.79
##                          0.14                          1.87
##                         -0.34                          0.61
##                         -1.10                         -0.46
##                         -3.70                          0.00
##                         -1.44                         -0.15
##                          0.95                          1.98
```

Table 3 provides a summary of the student calibrations. When there is a relatively large person sample size, it may be more useful to present the results as they relate to individual persons or subsets of the person sample as they are relevant to the purpose of the analysis.

```
# Calculate average ratings for students:
Person_Avg_Rating <- apply(ratings, 1, mean)

# Combine person calibration results in a table:
RS_MFRM_Table3 <- cbind.data.frame(rownames(student.
    locations_RSMFR),
                          Person_Avg_Rating,
                          student.locations_RSMFR$
                              theta,
                          student.locations_RSMFR$se,
                          student.fit$outfitPerson,
                          student.fit$outfitPerson_t,
```

```
                              student.fit$infitPerson,
                              student.fit$infitPerson_t)

names(RS_MFRM_Table3) <- c("Student ID", "Average
   Rating", "Student Location","Student SE","Outfit
   MSE","Std. Outfit", "Infit MSE","Std. Infit")

# Sort Table 3 by student location:
RS_MFRM_Table3 <- RS_MFRM_Table3[order(-RS_MFRM_
   Table3$`Student Location`),]

# Round the numeric values (all columns except the
   first one) to 2 digits:
RS_MFRM_Table3[, -1] <- round(RS_MFRM_Table3[,-1],
   digits = 2)

# Print the first six rows of the table to the
   console
print.data.frame(RS_MFRM_Table3[1:6,], row.names =
   FALSE)
```

```
##   Student ID Average Rating Student Location
##           69           3.00             7.60
##           44           2.95             6.57
##           45           2.95             6.57
##           12           2.95             6.44
##           70           2.95             6.44
##          269           2.95             6.44
##   Student SE Outfit MSE Std. Outfit Infit MSE
##         1.47       0.02       -0.81      0.03
##         0.87       0.54       -0.36      0.71
##         0.87       0.44       -0.54      0.68
##         0.87       0.54       -0.36      0.71
##         0.87       0.76       -0.01      0.79
##         0.87       0.54       -0.36      0.71
##   Std. Infit
##        -1.02
##        -0.18
##        -0.23
##        -0.18
##        -0.06
##        -0.18
```

Table 4 provides a summary of the student calibrations within subgroups.

```
# Calculate average ratings for student subgroups:
group.1.style <- subset(style, style$language == 1)
group.2.style <- subset(style, style$language == 2)

group.1_Avg_Rating <- mean(apply(group.1.style[, -c
    (1:2)], 1, mean))
group.2_Avg_Rating <- mean(apply(group.2.style[, -c
    (1:2)], 1, mean))

Subgroup_Avg_Rating <- c(group.1_Avg_Rating, group.2_
    Avg_Rating)

# Combine subgroup calibration results in a table:
RS_MFRM_Table4 <- cbind.data.frame(subgroup.estimates
    $parameter,
                        Subgroup_Avg_Rating,
                        subgroup.estimates$xsi,
                        subgroup.estimates$se.xsi,
                        c(mean(fit_group_1$
                            outfitPerson), mean(fit_
                            group_2$outfitPerson)),
                        c(mean(fit_group_1$
                            outfitPerson_t), mean(
                            fit_group_2$outfitPerson
                            _t)),
                        c(mean(fit_group_1$
                            infitPerson), mean(fit_
                            group_2$infitPerson)),
                        c(mean(fit_group_1$
                            infitPerson_t), mean(fit
                            _group_2$infitPerson_t))
                        )


names(RS_MFRM_Table4) <- c("Subgroup", "Average
    Rating", "Subgroup Location","Subgroup Location SE
    ","Outfit MSE","Std. Outfit", "Infit MSE","Std.
    Infit")

# Sort Table 4 by subgroup location:
RS_MFRM_Table4 <- RS_MFRM_Table4[order(-RS_MFRM_
    Table4$`Subgroup Location`),]
```

```
# Round the numeric values (all columns except the
    first one) to 2 digits:
RS_MFRM_Table4[, -1] <- round(RS_MFRM_Table4[,-1],
    digits = 2)

# Print the table to the console
print.data.frame(RS_MFRM_Table4, row.names = FALSE)

##    Subgroup Average Rating Subgroup Location
##    language2           1.59              0.07
##    language1           1.63             -0.07
##    Subgroup Location SE Outfit MSE Std. Outfit
##                    0.02        0.93       -0.24
##                    0.02        0.85       -0.41
##    Infit MSE Std. Infit
##         0.94      -0.21
##         0.86      -0.41
```

6.2   Another Example: Running PC-MFRM with Long-Format Data using the *TAM* Package

In the next section, we provide a step-by-step demonstration of a MFRM analysis using the *TAM* package (Robitzsch et al., 2021) for data that are stored in *long format*. We encourage readers to use the example data set for this chapter that is provided in the online supplement to conduct the analysis along with us.

For this example, we use a subset of the writing assessment data that includes students' scores related to four domains: Style, Organization, Conventions, and Sentence Formation.

Compared to the first example in this chapter, our description of the second example is less detailed. In cases where there are important differences between the two examples, we describe them. In other cases, we encourage readers to refer to the first example.

Prepare for the Analyses

Before proceeding with the analysis, readers should ensure that they have installed and loaded all three of the packages described earlier in this chapter: *TAM*, *WrightMap*, and *psych*.

Next, we will import the data for our analysis. The data for this example are stored in the file named `writing.csv`. We will save these data in an object called `writing`.

```
writing <- read.csv("writing.csv")
```

Note that the writing data are in *long* format: This means that there are multiple rows for each element within the object of measurement. In the case of our example data, there are multiple rows for each student. Each row includes one rater's ratings of one student on all four of the domains in the assessment: Style, Organization, Conventions, and Sentence Formation. We can see this structure by printing the first six rows of the data.frame object to our console using the head() function.

```
head(writing)
```

```
##   student language rater style org conv sent_form
## 1       3        1     1     2   2    2         2
## 2       3        1     2     2   2    2         3
## 3       3        1     3     2   2    1         2
## 4       3        1     4     2   3    2         3
## 5       3        1     5     3   3    2         3
## 6       3        1     6     3   3    2         2
```

Next, we will explore the data using descriptive statistics using the `summary()` function.

```
summary(writing)
```

```
##      student          language          rater
##  Min.   :   3.0   Min.   :1.000   Min.   : 1
##  1st Qu.: 392.0   1st Qu.:1.000   1st Qu.: 6
##  Median : 782.0   Median :2.000   Median :11
##  Mean   : 784.3   Mean   :1.532   Mean   :11
##  3rd Qu.:1171.2   3rd Qu.:2.000   3rd Qu.:16
##  Max.   :1574.0   Max.   :2.000   Max.   :21
##      style            org             conv
##  Min.   :0.000   Min.   :0.000   Min.   :0.000
##  1st Qu.:1.000   1st Qu.:1.000   1st Qu.:1.000
##  Median :2.000   Median :2.000   Median :2.000
##  Mean   :1.613   Mean   :1.521   Mean   :1.711
##  3rd Qu.:2.000   3rd Qu.:2.000   3rd Qu.:2.000
##  Max.   :3.000   Max.   :3.000   Max.   :3.000
##    sent_form
##  Min.   :0.000
##  1st Qu.:1.000
##  Median :2.000
```

```
##   Mean    :1.918
##   3rd Qu.:3.000
##   Max.    :3.000
```

From the summary of `writing`, we can see there are no missing data. In addition, we can get a general sense of the scales, range, and distribution of each variable in the data set. First, we can see that student identification numbers range from 3 to 1574. We can identify the number of unique students in the data using the following code.

```
length(unique(writing$student))
```

```
## [1] 372
```

There are 372 unique student identification numbers in our data. Returning to the summary of the writing data, we can see that the minimum rating on each domain was $x = 0$, and the maximum rating was $x = 3$.

6.2.1   Specify the PC-MFRM:

We will analyze the writing data using a Partial Credit MFRM (PC-MFRM), where we specify rating scale category thresholds separately for the domains in the writing assessment. The facets in this model will include raters and domains.

$$\ln\left(\frac{P_{n\,\mathrm{mi}(x=k)}}{P_{n\,\mathrm{mi}(x=k-1)}}\right) = \theta_n - \delta_m - \lambda_i - \tau_{mk} \tag{6.3}$$

In Equation 6.3, θ_n and λ_i are defined as in Equation 6.1 and Equation 6.2. δ_m is the logit-scale location for domain m. Lower domain locations indicate relatively easy domains, and higher domain locations indicate relatively difficult domains. τ_{mk} is the rating scale category threshold where there is an equal probability for a rating in category k and category $k - 1$, specific to domain m. This formulation of the PC-MFR model allows us to examine each rater's use of the rating scale separately.

With long format data, we need to ensure that the person identification numbers (in this case, student labels) are sorted from low to high before we can run the analysis.

```
writing <- writing[order(writing$student), ]
```

Next, we specify the components of the PC-MFRM. In this long-format data analysis, the *TAM* package will treat domains as "items" because they make up the columns of the response matrix. Therefore, our `writing.facets` object includes raters.

```
writing.facets <- writing[, c("rater"), drop = FALSE]
```

Next, we identify the object of measurement as students.

```
writing.pid <- writing$student
```

The response matrix includes students' ratings on the four domains.

```
writing.resp <- subset(writing, select = -c(student,
    language, rater))
```

We specify the PC-MFRM from Equation 6.3 in an object for use with *TAM* as follows. First, we specify a name for the model object (`writing_PC_MFRM`), which is defined using the tilde symbol (~), followed by the facet names. As a reminder, the model must include a facet named `item`; in our example, the item facet is made up of domains, because the domains make up the columns in our long-format response matrix. We also include `rater` as a facet. Finally, we use `item:step` to indicate the PC model.

```
PC.writing.formula <- ~ item + rater + item:step
```

We run the PC-MFR model using the `tam.mml.mfr()` function:

```
writing_PC_MFRM.model <- tam.mml.mfr(resp=writing.
    resp, facets=writing.facets,
                      formulaA=PC.writing.
                      formula, pid=writing.
                      pid, verbose = FALSE,
                      constraint = "items"
                      )
```

Analysts who are interested in overall model-fit indices or other details that are included in the summary of the MFRM may request it using the `summary()` function.

```
summary(writing_PC_MFRM.model)
```

Facet Results

Because we used a PC formulation of the MFRM, the thresholds were estimated separately for each rater. We save the facet estimates for domains and raters as we did in the first example. However, for the threshold estimates, we extract the values labeled `item:step`.

```
# Save the facet estimates:
facet.estimates <- writing_PC_MFRM.model$xsi.facets #
    all facets together
```

```
# Extract results for each facet separately:
domain.estimates <- subset(facet.estimates, facet.
    estimates$facet == "item")
rater.estimates <- subset(facet.estimates, facet.
    estimates$facet == "rater")

# Extract domain-specific threshold estimates:
threshold.estimates <- subset(facet.estimates, facet.
    estimates$facet == "item:step")
```

Domain facet results

Next, we will examine the results for the domain facet. Because there are only four domains, we can print the estimates to the console to view them. We can also calculate summary statistics for the domain facet estimates using the summary() function.

```
domain.estimates
```

```
##   parameter facet        xsi      se.xsi
## 1     style  item  0.26265320  0.01585224
## 2       org  item  0.63426014  0.01584412
## 3      conv  item -0.08840333  0.01599162
## 4 sent_form  item -0.80851000  0.02753292
```

```
summary(domain.estimates)
```

```
##   parameter               facet
##   Length:4            Length:4
##   Class :character    Class :character
##   Mode  :character    Mode  :character
##
##
##
##        xsi                  se.xsi
##   Min.   :-0.80851    Min.   :0.01584
##   1st Qu.:-0.26843    1st Qu.:0.01585
##   Median : 0.08712    Median :0.01592
##   Mean   : 0.00000    Mean   :0.01881
##   3rd Qu.: 0.35555    3rd Qu.:0.01888
##   Max.   : 0.63426    Max.   :0.02753
```

From the summary of domain locations, we can see that the domain difficulty estimates range from $\delta = -0.81$ logits for the easiest domain to $\delta = 0.63$ for the most difficult domain.

Rater Facet Results

Nest, we will examine the rater estimates using the summary() function.

```
summary(rater.estimates)
```

```
##    parameter              facet
##  Length:21          Length:21
##  Class :character   Class :character
##  Mode  :character   Mode  :character
##
##
##
##        xsi                  se.xsi
##  Min.   :-0.549431   Min.   :0.03623
##  1st Qu.:-0.191938   1st Qu.:0.03625
##  Median :-0.001687   Median :0.03628
##  Mean   : 0.000000   Mean   :0.04231
##  3rd Qu.: 0.164105   3rd Qu.:0.03635
##  Max.   : 0.764278   Max.   :0.16236
```

The rater location estimates from this model range from $\lambda = -0.55$ logits for the most lenient rater to $\lambda = 0.76$ logits for the most severe rater.

Student Facet Results

Next, we will estimate student locations and store the results in an object called student.ach. We will then store the student identification numbers, location estimates, and standard errors in a new object called student. locations_PCMFR.

```
student.ach <- tam.wle(writing_PC_MFRM.model,
    progress = FALSE)

student.locations_PCMFR <- cbind.data.frame(student.
    ach$pid, student.ach$theta, student.ach$error)

names(student.locations_PCMFR) <- c("id", "theta", "
    se")
```

Next, we will examine the student location estimates using the summary() function.

```
summary(student.locations_PCMFR$theta)
```

```
##     Min. 1st Qu.  Median    Mean 3rd Qu.    Max.
##  -7.1284 -1.4232  0.9448  0.7117  2.9680  7.6930
```

Student achievement ranges from $\theta = -7.13$ logits for the student with the lowest achievement estimate to $\theta = 7.69$ logits for the student with the highest achievement estimate. On average, the students were located higher ($M\ \theta = 0.71$ logits) than the average domain location ($M\ \delta = 0.00$).

Threshold estimates

Because we used a PC formulation of the MFRM, we have separate estimates of the rating scale category thresholds for each domain. As a result, we can evaluate rating scale category thresholds from the perspective of rating scale analysis (Linacre, 2002). For example, we can examine the degree to which the thresholds are non-decreasing across increasing rating scale categories, and the degree to which each category describes a unique range of locations on the latent variable. In the following code, we organize the threshold values in a matrix with separate rows for each domain.

```
n.domains <- nrow(domain.estimates)
n.thresholds <- max(writing.resp)

domain_taus <- matrix(data = NA, nrow = n.domains,
    ncol = n.thresholds)

for(domain.number in 1:n.domains){

  domain.threshold.labels <- NULL
  for(step.number in 1:n.thresholds){
    domain.threshold.labels[step.number] <- paste(
        domain.estimates$parameter[domain.number], ":
        step", step.number, sep = "")
  }

  domain.thresholds <- subset(threshold.estimates,
                         threshold.estimates$
                             parameter %in% domain.
                             threshold.labels)

  domain.thresholds.t <- t(domain.thresholds$xsi)

  domain_taus[domain.number,] <- domain.thresholds.t
      [1,]
}

domain_taus <- cbind.data.frame(c(1:n.domains),
    domain_taus)
names(domain_taus) <- c("domain", "t1", "t2", "t3")
```

```
domain_taus
```

```
##    domain         t1           t2         t3
## 1       1  -3.725268   0.27603490   3.449234
## 2       2  -3.642835   0.20323183   3.439604
## 3       3  -3.610870  -0.08923873   3.700108
## 4       4  -3.124334  -0.20954280   3.333877
```

Wright Map

Next, we will plot a Wright Map to display the locations of the parameter estimates for our PC-MFR model. The `wrightMap()` function requires us to store the domain location estimates as a matrix that shows domain-specific threshold locations. We already did this when we created the `domain_taus` object earlier.

We can plot the Wright map using the `wrightMap()` function. We specify several graphical parameters to modify the appearance of the plot.

```
wrightMap(thetas =cbind(student.locations_PCMFR$theta
       ,
                         rater.estimates$xsi),
          axis.persons = "",
          dim.names = c("Students", "Raters"),
          thresholds = domain_taus[,-1],
          show.thr.lab  = TRUE,
          label.items.rows= 2,
          label.items = domain.estimates$parameter,
          axis.items = "Domains",
          main.title = "Partial Credit Many-Facet
              Rasch Model \nWright Map: Style Ratings"
              ,
          cex.main = .6)
```

Partial Credit Many–Facet Rasch Model
Wright Map: Style Ratings

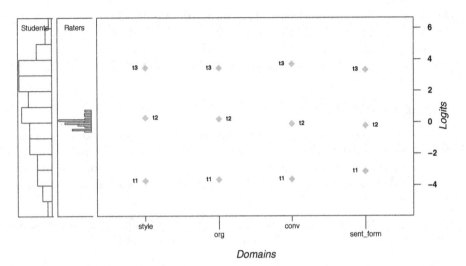

Domains

```
##                 t1           t2          t3
## [1,]    -3.725268   0.27603490    3.449234
## [2,]    -3.642835   0.20323183    3.439604
## [3,]    -3.610870  -0.08923873    3.700108
## [4,]    -3.124334  -0.20954280    3.333877
```

In this *Wright Map* display, the results from the PC-MFRM analysis of the writing assessment ratings are summarized graphically using the same format as the first example in this chapter. The major differences between the models are that domains are included as a facet instead of student subgroups, and each domain has a unique set of threshold estimates.

Evaluate Model-Data Fit

Next, we will evaluate model-data fit using the same procedures as described earlier in this chapter.

Unidimensionality

First, we will construct a residual matrix.

```
resids <- IRT.residuals(writing_PC_MFRM.model)

# Extract the raw residuals from the residuals object
:
resid.matrix <- as.data.frame(resids$residuals)
```

Next, we will calculate standardized residuals and save them in a matrix.

```
std.resid.matrix <- as.data.frame(resids$stand_
    residuals)
```

Next, we will calculate the variance in observations due to Rasch-model-estimated locations:

```
# Variance of the observations: VO
observations.vector <- as.vector(as.matrix(writing.
    resp))
VO <- var(observations.vector)
```

```
# Variance of the residuals: VR
residuals.vector <- as.vector(as.matrix(resid.matrix)
    )
VR <- var(residuals.vector)
```

```
# Raw variance explained by Rasch measures: (VO - VR)
    / VO
(VO - VR)/VO
```

```
## [1] 0.7143585
```

```
# Express the result as a percent:
((VO - VR)/VO) * 100
```

```
## [1] 71.43585
```

Approximately 71.44% of the variance in ratings can be explained by the PC-MFRM estimates of student, domain, and rater locations on the logit scale that represents the latent variable.

Principal Components Analysis of Standardized Residual Correlations

Next, we will evaluate the MFRM requirement for unidimensionality using a PCA of standardized residual correlations.

```
pca <- pca(as.matrix(std.resid.matrix), rotate = "
    none")
```

```
contrasts <- c(pca$values[1], pca$values[2], pca$
    values[3], pca$values[4], pca$values[5])
```

```
plot(contrasts, ylab = "Eigenvalues for Contrasts",
    xlab = "Contrast Number", main = "Contrasts from
    PCA Standardized Residual Correlations \n(PC-MFRM)
    ", cex.main = .8)
```

**Contrasts from PCA of Standardized Residual Correlations
(PC–MFRM)**

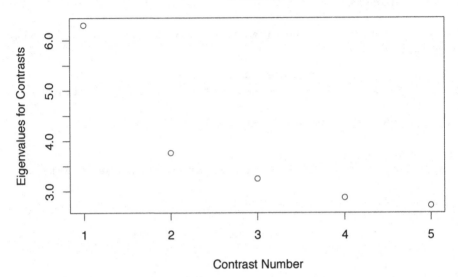

Contrast Number

In this example, there are three contrasts that have an eigenvalue larger than Linacre's (2016)'s critical value of 2.00. This result suggests potential multi-dimensionality. We will explore the results further using residual-based fit statistics.

Summaries of Residuals: Infit & Outfit Statistics

Next, we will evaluate model-data fit for individual elements of our facets (students, subgroups, and raters) using numeric summaries of the residuals associated with each element, as we have done in previous chapters.

Student Fit

First, we will examine student fit using numeric outfit and infit statistics. We can request these statistics for each student using the `tam.personfit()` function. We will store the student fit results in an object called `student.fit`, and then request a summary of the results.

```
student.fit <- tam.personfit(writing_PC_MFRM.model)
summary(student.fit)
```

```
##    outfitPerson        outfitPerson_t
##   Min.   :0.2405     Min.   :-3.7371
##   1st Qu.:0.8048     1st Qu.:-1.2204
##   Median :0.9465     Median :-0.2436
##   Mean   :0.9618     Mean   :-0.2482
##   3rd Qu.:1.0954     3rd Qu.: 0.7088
##   Max.   :1.9949     Max.   : 5.3917
```

```
##       infitPerson          infitPerson_t
##    Min.    :0.4884      Min.     :-3.9688
##    1st Qu.:0.8268      1st Qu.:-1.1888
##    Median :0.9620      Median :-0.1598
##    Mean    :0.9714      Mean     :-0.2213
##    3rd Qu.:1.0871      3rd Qu.: 0.6185
##    Max.    :1.9824      Max.    : 5.4138
```

On average, the outfit and infit MSE statistics are slightly lower than the expected value of 1 (M outfit = 0.96, M infit = 0.97). The average values of the standardized fit statistics are also slightly lower than their expected value of 0 (M std. outfit = -0.25, M std. infit = -0.22). For both the standardized and unstandardized fit statistics, there is notable variability across the student sample. This result suggests that model-data fit varies for individual students.

Domain and Rater Fit

We can also examine model-data fit related to the domain and rater facets. In the *TAM* package, fit analysis for facets besides the object of measurement uses combinations of elements within facets. In our example, fit statistics are calculated for rater*subgroup combinations.

```
rater.domain.fit <- msq.itemfit(writing_PC_MFRM.model
     )
rater.domain.fit <- rater.domain.fit$itemfit
summary(rater.domain.fit)
```

```
##        item                fitgroup
##    Length:84          Min.     : 1.00
##    Class :character    1st Qu.:21.75
##    Mode   :character    Median :42.50
##                          Mean    :42.50
##                          3rd Qu.:63.25
##                          Max.    :84.00
##        Outfit              Outfit_t
##    Min.    :0.5309      Min.     :-7.3816
##    1st Qu.:0.8521      1st Qu.:-1.9994
##    Median :0.9584      Median :-0.5210
##    Mean    :0.9773      Mean     :-0.4319
##    3rd Qu.:1.1410      3rd Qu.: 1.5425
##    Max.    :1.5076      Max.    : 5.7335
##        Outfit_p              Infit
##    Min.    :0.000000    Min.     :0.5627
##    1st Qu.:0.003517    1st Qu.:0.8657
##    Median :0.092685    Median :0.9784
##    Mean    :0.231879    Mean     :0.9813
##    3rd Qu.:0.456375    3rd Qu.:1.1088
```

```
##    Max.    :0.990028   Max.    :1.5218
##       Infit_t             Infit_p
##    Min.    :-7.4049   Min.    :0.000000
##    1st Qu.:-1.9621   1st Qu.:0.002148
##    Median :-0.2818   Median :0.103798
##    Mean    :-0.4201   Mean    :0.234087
##    3rd Qu.: 1.5044   3rd Qu.:0.417554
##    Max.    : 6.3830   Max.    :0.981096
```

As needed, researchers can also examine model-data fit statistics specific to levels of explanatory facets, such as domains in the current example. This can be accomplished by calculating summaries of fit statistics within domains. The code below calculates averages of rater fit within each domain. For a more detailed view of rater fit within domains, we create an object with domain-specific rater fit statistics in the subsequent code block.

```
ngroups <- nrow(domain.estimates)

domain.fit <- matrix(data = NA, nrow = ncol(domain.
    estimates), ncol = 4 )

for(domain.number in 1:ncol(domain.estimates)){

  d <- domain.estimates$parameter[domain.number]

  rater.domain.labels <- NULL

    for(r in 1:nrow(rater.estimates)){
      rater_label <- rater.estimates$parameter[r]
      rater.domain.labels[r] <- paste(d, "-", rater_
        label, sep = "")
    }

  rater.domain.fit.subset <- subset(rater.domain.fit,
      rater.domain.fit$item %in% rater.domain.labels)

  # add the fit statistics to the matrix:
  domain.fit[domain.number, ] <-   c(mean(rater.domain
    .fit.subset$Infit),
                                 mean(rater.domain.
                                   fit.subset$
                                   Outfit),
                                 mean(rater.domain.
                                   fit.subset$
                                   Infit_t),
                                 mean(rater.domain.
```

```
                                              fit.subset$
                                              Outfit_t))
  }
```

 # add domain labels to the domain.fit object:

```
domain.fit <- cbind.data.frame(domain.estimates$
   parameter,
                               domain.fit)
```

 # Convert the domain fit results to a data frame
   object and add meaningful column names:
```
domain.fit_results <- as.data.frame(domain.fit)
```

```
names(domain.fit_results) <- c("domain", "Mean_Infit_
   MSE", "Mean_Outfit_MSE",
                                "Mean_Std_Infit", "
                                Mean_Std_Outfit")
```

Now that we have created a data frame with the domain-specific (average) fit statistics, we can summarize the results.

```
summary(domain.fit_results)
```

```
##       domain          Mean_Infit_MSE
##   Length:4            Min.   :0.9424
##   Class :character    1st Qu.:0.9581
##   Mode  :character    Median :0.9706
##                       Mean   :0.9813
##                       3rd Qu.:0.9938
##                       Max.   :1.0415
##   Mean_Outfit_MSE     Mean_Std_Infit
##   Min.   :0.9390      Min.   :-1.0223
##   1st Qu.:0.9488      1st Qu.:-0.7369
##   Median :0.9567      Median :-0.5483
##   Mean   :0.9773      Mean   :-0.4201
##   3rd Qu.:0.9852      3rd Qu.:-0.2315
##   Max.   :1.0566      Max.   : 0.4386
##   Mean_Std_Outfit
##   Min.   :-1.0032
##   1st Qu.:-0.7373
##   Median :-0.6383
##   Mean   :-0.4319
##   3rd Qu.:-0.3329
##   Max.   : 0.5522
```

Finally, it may be useful to examine fit statistics as they apply to individual raters. This can be accomplished by extracting rater-specific fit statistics within each domain. The code below calculates rater fit statistics within each subgroup.

```
n.domains <- nrow(domain.estimates)
n.raters <- nrow(rater.estimates)

rater.fit <- matrix(data = NA, nrow = n.raters, ncol
    = (n.domains * 4) + 1 )

for(rater.number in 1:nrow(rater.estimates)){

    if(rater.number < 10) r <- paste("rater_", rater.
        number, "_", sep = "")
    if(rater.number >= 10) r <- paste("rater_", rater.
        number, sep = "")

    rater.domain.labels <- NULL

    for(d in 1:nrow(domain.estimates)){
        domain_label <- domain.estimates$parameter[d]
        rater.domain.labels[d] <- paste(domain_label, "-"
            , r, sep = "")
    }

    rater.domain.fit.subset <- subset(rater.domain.fit,
        rater.domain.fit$item %in% rater.domain.labels)

    # calculate rater-specific fit statistics:
    rater.outfit <- rater.domain.fit.subset$Outfit
    rater.infit <- rater.domain.fit.subset$Outfit
    rater.std.outfit <- rater.domain.fit.subset$Outfit_
        t
    rater.std.infit <- rater.domain.fit.subset$Infit_t

    # add the fit statistics to the matrix:
    rater.fit[rater.number, ] <-  c(rater.number, rater
        .outfit, rater.infit,
                                    rater.std.outfit,
                                    rater.std.infit)
}


# Convert the rater fit results to a dataframe object
    and add meaningful column names:
```

```
rater.fit_results <- as.data.frame(rater.fit)

infit_mse_labels <- NULL
for(domain in 1:n.domains){
  d <- domain.estimates$parameter[domain]
  infit_mse_labels[domain] <- paste("Infit_MSE_", d,
    sep = "")
}

outfit_mse_labels <- NULL
for(domain in 1:n.domains){
  d <- domain.estimates$parameter[domain]
  outfit_mse_labels[domain] <- paste("outfit_MSE_", d
    , sep = "")
}

std_infit_mse_labels <- NULL
for(domain in 1:n.domains){
  d <- domain.estimates$parameter[domain]
  std_infit_mse_labels[domain] <- paste("std_infit_
    MSE_", d, sep = "")
}

std_outfit_mse_labels <- NULL
for(domain in 1:n.domains){
  d <- domain.estimates$parameter[domain]
  std_outfit_mse_labels[domain] <- paste("std_outfit_
    MSE_", d, sep = "")
}

names(rater.fit_results) <- c("Rater", outfit_mse_
  labels, infit_mse_labels,
                              std_outfit_mse_labels,
                              std_infit_mse_labels
                              )
```

Now that we have created a data frame with the rater-specific fit statistics, we can summarize the results.

```
summary(rater.fit_results)
```

```
##       Rater        outfit_MSE_style  outfit_MSE_org
##   Min.    : 1   Min.     :0.6203   Min.    :0.5349
##   1st Qu.: 6   1st Qu.:0.9423     1st Qu.:0.8183
##   Median :11   Median :0.9803     Median :0.9543
```

```
##   Mean    :11    Mean     :1.0566    Mean     :0.9614
##   3rd Qu.:16    3rd Qu.:1.1949    3rd Qu.:1.1284
##   Max.    :21    Max.     :1.5076    Max.     :1.2498
##   outfit_MSE_conv    outfit_MSE_sent_form
##   Min.    :0.5314    Min.     :0.5309
##   1st Qu.:0.7423    1st Qu.:0.8777
##   Median :0.9431    Median :0.9336
##   Mean    :0.9390    Mean     :0.9520
##   3rd Qu.:1.1079    3rd Qu.:1.0824
##   Max.    :1.2430    Max.     :1.2808
##   Infit_MSE_style    Infit_MSE_org
##   Min.    :0.6203    Min.     :0.5349
##   1st Qu.:0.9423    1st Qu.:0.8183
##   Median :0.9803    Median :0.9543
##   Mean    :1.0566    Mean     :0.9614
##   3rd Qu.:1.1949    3rd Qu.:1.1284
##   Max.    :1.5076    Max.     :1.2498
##   Infit_MSE_conv    Infit_MSE_sent_form
##   Min.    :0.5314    Min.     :0.5309
##   1st Qu.:0.7423    1st Qu.:0.8777
##   Median :0.9431    Median :0.9336
##   Mean    :0.9390    Mean     :0.9520
##   3rd Qu.:1.1079    3rd Qu.:1.0824
##   Max.    :1.2430    Max.     :1.2808
##   std_outfit_MSE_style std_outfit_MSE_org
##   Min.    :-5.6828    Min.      :-7.2549
##   1st Qu.:-0.6897    1st Qu.:-2.3590
##   Median :-0.2269    Median :-0.5642
##   Mean    : 0.5522    Mean     :-0.6280
##   3rd Qu.: 2.4048    3rd Qu.: 1.4881
##   Max.    : 5.7335    Max.     : 2.9812
##   std_outfit_MSE_conv std_outfit_MSE_sent_form
##   Min.    :-7.3816    Min.     :-5.8979
##   1st Qu.:-3.6658    1st Qu.:-1.3706
##   Median :-0.7047    Median :-0.6790
##   Mean    :-1.0032    Mean     :-0.6487
##   3rd Qu.: 1.3936    3rd Qu.: 0.9099
##   Max.    : 2.9876    Max.     : 2.9067
##   std_infit_MSE_style std_infit_MSE_org
##   Min.    :-5.73380    Min.     :-7.2672
##   1st Qu.:-0.99401    1st Qu.:-2.2358
##   Median :-0.06115    Median :-0.1441
##   Mean    : 0.43862    Mean     :-0.4549
##   3rd Qu.: 2.47375    3rd Qu.: 1.2517
##   Max.    : 6.38299    Max.     : 3.8177
```

```
## std_infit_MSE_conv std_infit_MSE_sent_form
## Min.    :-7.4049    Min.    :-6.5507
## 1st Qu.:-4.1189    1st Qu.:-1.4860
## Median :-0.6799    Median :-0.7798
## Mean    :-1.0223    Mean    :-0.6418
## 3rd Qu.: 1.4964    3rd Qu.: 0.8452
## Max.    : 3.8886    Max.    : 4.0513
```

Graphical Displays of Residuals

Next, we will construct plots of standardized residuals associated with individual raters within each domain.

```
# Before constructing the plots, find the maximum and
    minimum values of the standardized residuals to
    set limits for the axes:
max.resid <- ceiling(max(std.resid.matrix))
min.resid <- ceiling(min(std.resid.matrix))

# The code below will produce plots of standardized
    residuals for selected raters as listed in raters.
    to.plot:
raters.to.plot <- c(1:2)

for(rater.number in raters.to.plot){

  if(rater.number < 10) r <- paste("rater_", rater.
    number, "_", sep = "")
  if(rater.number >= 10) r <- paste("rater_", rater.
    number, sep = "")

  rater.domain.labels <- NULL

  for(d in 1:nrow(domain.estimates)){
      domain_label <- domain.estimates$parameter[d]
    rater.domain.labels[d] <- paste(domain_label, "-"
      , r, sep = "")
    }

  std.resid.subset <- subset(resids$stand_residuals,
    select = rater.domain.labels)

  for(domain.number in 1:n.domains){
    domain.name <- domain.estimates$parameter[domain.
      number]
```

```
plot(std.resid.subset[, domain.number], ylim = c(
    min.resid, max.resid),
    main = paste("Standardized Residuals for Rater
        ", rater.number, " Domain ", domain.name,
        sep = ""),
    ylab = "Standardized Residual", xlab = "Person
        Index")
abline(h = 0, col = "blue")
abline(h=2, lty = 2, col = "red")
abline(h=-2, lty = 2, col = "red")

legend("topright", c("Std. Residual", "Observed
    = Expected", "+/- SD"), pch = c(1, NA, NA),
    lty = c(NA, 1, 2),
    col = c("black", "blue", "red"), cex =
        .8)
}
}
```

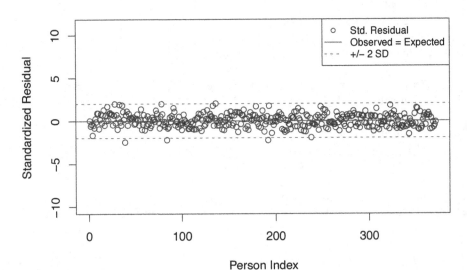

Standardized Residuals for Rater 1 Domain = style

Standardized Residuals for Rater 1 Domain = org

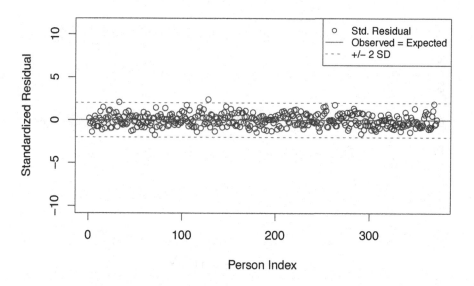

Standardized Residuals for Rater 1 Domain = conv

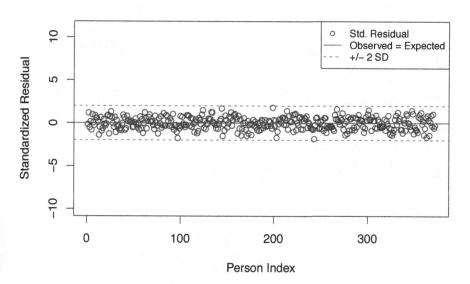

Standardized Residuals for Rater 1 Domain = sent_form

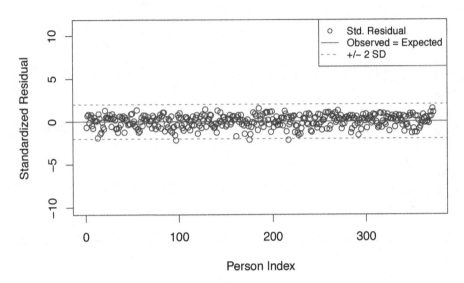

Standardized Residuals for Rater 2 Domain = style

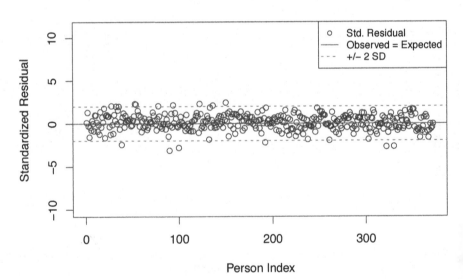

Standardized Residuals for Rater 2 Domain = org

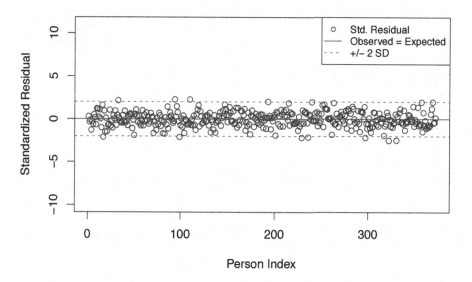

Standardized Residuals for Rater 2 Domain = conv

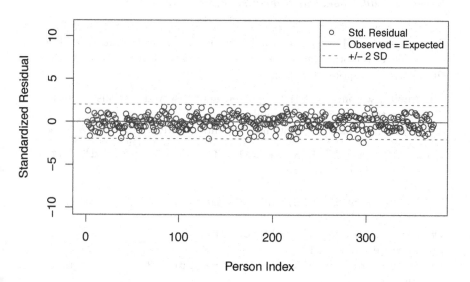

Standardized Residuals for Rater 2 Domain = sent_form

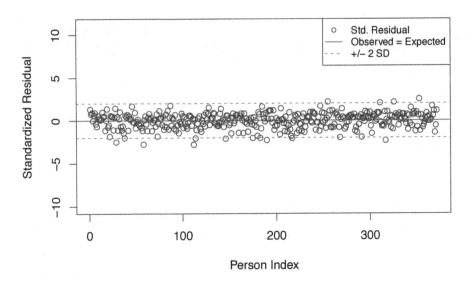

A separate plot is produced for each rater*domain combination.

Expected and Observed Response Functions

Finally, we will construct plots of expected and observed response functions. By default, the *TAM* package combines the item facet (in this case, domains), with levels of the other facets (in this case, raters) when constructing expected and observed response function plots.

For brevity, we only plot the expected and observed response functions for three selected rater*domain combinations. Readers can adjust the `items=` specification to construct plots for elements of interest for their analyses.

```
plot(writing_PC_MFRM.model, type = "expected", items
    = c(1:3))
```

```
## Iteration in WLE/MLE estimation   1   | Maximal
       change   2.2459
## Iteration in WLE/MLE estimation   2   | Maximal
       change   0.8781
## Iteration in WLE/MLE estimation   3   | Maximal
       change   0.5385
## Iteration in WLE/MLE estimation   4   | Maximal
       change   0.1794
## Iteration in WLE/MLE estimation   5   | Maximal
       change   0.0144
## Iteration in WLE/MLE estimation   6   | Maximal
       change   1e-04
```

```
## Iteration in WLE/MLE estimation   7    | Maximal
   change   0
## ----
##   WLE Reliability= 0.993
```

Expected Scores Curve – Item style–rater_1_

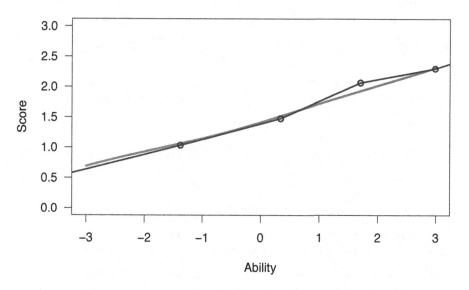

Expected Scores Curve – Item style–rater_18

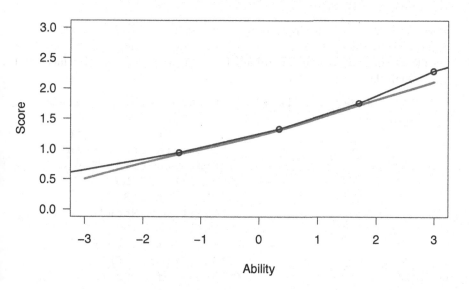

Expected Scores Curve – Item style–rater_19

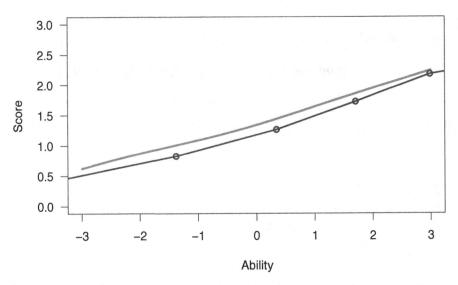

Summarize the Results in Tables

As a final step, we will create tables that summarize the calibrations of the students, domains, raters, and rating scale category thresholds.

Table 1 provides an overview of the logit-scale locations, standard errors and fit statistics for all of the facets in the analysis. This table provides a succinct overview of the location estimates and numeric model-data fit statistics for the facets in a MFRM.

Because of the estimation procedure for the MFRM in *TAM*, fit statistics are combined for the item facet (in this case, domains) and other facets. As a result, the fit statistics in this table will be the same for the rater facet and the domain facets.

```
PC_MFRM_summary.table.statistics <- c("Logit Scale
    Location Mean",
                        "Logit Scale Location SD",
                        "Standard Error Mean",
                        "Standard Error SD",
                        "Outfit MSE Mean",
                        "Outfit MSE SD",
                        "Infit MSE Mean",
                        "Infit MSE SD",
                        "Std. Outfit Mean",
                        "Std. Outfit SD",
                        "Std. Infit Mean",
                        "Std. Infit SD")
```

```
PC_MFRM_student.summary.results <- rbind(mean(student
    .locations_PCMFR$theta),
                            sd(student.locations_PCMFR
                                $theta),
                            mean(student.locations_
                                PCMFR$se),
                            sd(student.locations_PCMFR
                                $se),
                            mean(student.fit$
                                outfitPerson),
                            sd(student.fit$
                                outfitPerson),
                            mean(student.fit$
                                infitPerson),
                            sd(student.fit$infitPerson
                                ),
                            mean(student.fit$
                                outfitPerson_t),
                            sd(student.fit$
                                outfitPerson_t),
                            mean(student.fit$
                                infitPerson_t),
                            sd(student.fit$infitPerson
                                _t))


PC_MFRM_domain.summary.results <- rbind(mean(domain.
    estimates$xsi),
                            sd(domain.estimates$xsi),
                            mean(domain.estimates$se.
                                xsi),
                            sd(domain.estimates$se.xsi
                                ),
                            mean(rater.domain.fit$
                                Outfit),
                            sd(rater.domain.fit$Outfit
                                ),
                            mean(rater.domain.fit$
                                Infit),
                            sd(rater.domain.fit$Infit)
                                ,
                            mean(rater.domain.fit$
                                Outfit_t),
```

```
                                      sd(rater.domain.fit$Outfit
                                         _t),
                                      mean(rater.domain.fit$
                                         Infit_t),
                                      sd(rater.domain.fit$Infit_
                                         t))


PC_MFRM_rater.summary.results <- rbind(mean(rater.
   estimates$xsi),
                                      sd(rater.estimates$xsi),
                                      mean(rater.estimates$se.
                                         xsi),
                                      sd(rater.estimates$se.xsi)
                                         ,
                                      mean(rater.domain.fit$
                                         Outfit),
                                      sd(rater.domain.fit$Outfit
                                         ),
                                      mean(rater.domain.fit$
                                         Infit),
                                      sd(rater.domain.fit$Infit)
                                         ,
                                      mean(rater.domain.fit$
                                         Outfit_t),
                                      sd(rater.domain.fit$Outfit
                                         _t),
                                      mean(rater.domain.fit$
                                         Infit_t),
                                      sd(rater.domain.fit$Infit_
                                         t))



# Round the values for presentation in a table:
PC_MFRM_student.summary.results_rounded <- round(PC_
   MFRM_student.summary.results, digits = 2)

PC_MFRM_domain.summary.results_rounded <- round(PC_
   MFRM_domain.summary.results, digits = 2)

PC_MFRM_rater.summary.results_rounded <- round(PC_
   MFRM_rater.summary.results, digits = 2)


PC_MFRM_Table1 <- cbind.data.frame(PC_MFRM_summary.
   table.statistics,
```

```
                                     PC_MFRM_student.summary.
                                       results_rounded,
                                     PC_MFRM_domain.summary.
                                       results_rounded,
                                     PC_MFRM_rater.summary.
                                       results_rounded)

# add descriptive column labels:
names(PC_MFRM_Table1) <- c("Statistic", "Students", "
    Domains", "Raters")

# Print the table to the console
print.data.frame(PC_MFRM_Table1, row.names = FALSE)

##                       Statistic Students Domains
##   Logit Scale Location Mean         0.71    0.00
##      Logit Scale Location SD        2.92    0.61
##          Standard Error Mean        0.23    0.02
##            Standard Error SD        0.07    0.01
##             Outfit MSE Mean         0.96    0.98
##              Outfit MSE SD          0.23    0.20
##             Infit MSE Mean          0.97    0.98
##               Infit MSE SD          0.22    0.19
##          Std. Outfit Mean          -0.25   -0.43
##            Std. Outfit SD           1.47    2.66
##           Std. Infit Mean          -0.22   -0.42
##             Std. Infit SD           1.48    2.80
##   Raters
##      0.00
##      0.34
##      0.04
##      0.03
##      0.98
##      0.20
##      0.98
##      0.19
##     -0.43
##      2.66
##     -0.42
##      2.80
```

Table 2 summarizes the overall calibrations of individual raters. For data sets with manageable sample sizes such as the writing assessment example in this chapter, we recommend reporting details about each level of explanatory facets (e.g., individual raters) in a table similar to this one.

```
# Calculate the average rating for each rater:
n.raters <- nrow.(rater.estimates)

Avg_Rating_rater.domains <- NULL

for(rater.number in 1:n.raters){
  rater.subset <- subset(writing, writing$rater ==
    rater.number)
  rater.ratings <- as.vector(c(rater.subset$style,
                    rater.subset$org, rater.subset$
                      conv,
                    rater.subset$sent_form))
  Avg_Rating_rater.domains[rater.number] <- mean(
    rater.ratings)
}




# Combine rater calibration results in a table:
PC_MFRM_Table2 <- cbind.data.frame(c(1:nrow(rater.
  estimates)),
                        Avg_Rating_rater.domains,
                        rater.estimates$xsi,
                        rater.estimates$se.xsi,
                        rater.fit_results[, -1])

names(PC_MFRM_Table2) <- c("Rater ID", "Average
  Rating", "Rater Location", "Rater SE",
                        names(rater.fit_results[,
                          -1]))

# Sort Table 2 by rater severity:
PC_MFRM_Table2 <- PC_MFRM_Table2[order(-PC_MFRM_
  Table2$`Rater Location`),]

# Round the numeric values (all columns except the
  first one) to 2 digits:
PC_MFRM_Table2[, -1] <- round(PC_MFRM_Table2[,-1],
  digits = 2)

# Print the first six rows of the table to the
  console
print.data.frame(PC_MFRM_Table2[1:6,], row.names =
  FALSE)
```

```
## Rater ID Average Rating Rater Location Rater SE
##       21            1.70            0.76       0.16
##       10            1.81            0.50       0.04
##        8            1.62            0.41       0.04
##       20            1.65            0.27       0.04
##        6            1.81            0.17       0.04
##        5            1.76            0.16       0.04
## outfit_MSE_style outfit_MSE_org outfit_MSE_conv
##             1.51            1.25             1.03
##             1.19            1.23             0.94
##             0.95            0.82             0.92
##             1.06            0.97             1.08
##             1.17            0.91             0.89
##             1.23            1.05             1.23
## outfit_MSE_sent_form Infit_MSE_style
##                 1.14            1.51
##                 0.88            1.19
##                 1.00            0.95
##                 0.90            1.06
##                 0.92            1.17
##                 1.19            1.23
## Infit_MSE_org Infit_MSE_conv Infit_MSE_sent_form
##          1.25           1.03                1.14
##          1.23           0.94                0.88
##          0.82           0.92                1.00
##          0.97           1.08                0.90
##          0.91           0.89                0.92
##          1.05           1.23                1.19
## std_outfit_MSE_style std_outfit_MSE_org
##                 5.73               2.98
##                 2.33               2.78
##                -0.62              -2.36
##                 0.79              -0.34
##                 2.06              -1.24
##                 2.84               0.63
## std_outfit_MSE_conv std_outfit_MSE_sent_form
##                0.46                     1.53
##               -0.70                    -1.17
##               -1.03                     0.02
##                1.10                    -1.18
##               -1.39                    -0.70
##                2.81                     1.85
## std_infit_MSE_style std_infit_MSE_org
##                6.38              3.82
##                2.47              3.06
```

```
##                    -0.56                        -2.54
##                     0.21                         0.09
##                     2.29                        -0.83
##                     2.94                         1.25
##    std_infit_MSE_conv std_infit_MSE_sent_form
##                     0.67                         2.53
##                    -0.68                        -1.62
##                    -0.68                         0.85
##                     1.50                        -0.78
##                    -1.49                        -0.82
##                     3.05                         2.63
```

Table 3 summarizes the calibration of individual domains. For data sets with manageable sample sizes such as the writing assessment example in this chapter, we recommend reporting details about each element of explanatory facets (e.g., individual domains) in a table similar to this one.

```
# Calculate the average rating for each domain:
Avg_Rating_domains <- colMeans(writing.resp)

# Combine domain calibration results in a table:
PC_MFRM_Table3 <- cbind.data.frame(domain.estimates$
    parameter,
                        Avg_Rating_domains,
                        domain.estimates$xsi,
                        domain.estimates$se.xsi,
                        domain_taus[, -1],
                        domain.fit_results[, -1])

names(PC_MFRM_Table3) <- c("Domain", "Average Rating"
    , "Domain Location", "Domain SE",
                        "Threshold 1", "Threshold
                            2", "Threshold 3",
                        names(domain.fit_results
                            [, -1]))

# Sort Table 3 by domain difficulty:
PC_MFRM_Table3 <- PC_MFRM_Table3[order(-PC_MFRM_
    Table3$`Domain Location`),]

# Round the numeric values (all columns except the
    first one) to 2 digits:
PC_MFRM_Table3[, -1] <- round(PC_MFRM_Table3[,-1],
    digits = 2)

# Print the table to the console
```

```
print.data.frame(PC_MFRM_Table3, row.names = FALSE)
```

Domain	Average Rating	Domain Location
org	1.52	0.63
style	1.61	0.26
conv	1.71	-0.09
sent_form	1.92	-0.81

Domain SE	Threshold 1	Threshold 2	Threshold 3
0.02	-3.64	0.20	3.44
0.02	-3.73	0.28	3.45
0.02	-3.61	-0.09	3.70
0.03	-3.12	-0.21	3.33

Mean_Infit_MSE	Mean_Outfit_MSE	Mean_Std_Infit
0.98	0.96	-0.45
1.04	1.06	0.44
0.94	0.94	-1.02
0.96	0.95	-0.64

Mean_Std_Outfit
-0.63
0.55
-1.00
-0.65

Finally, Table 4 provides a summary of the student calibrations. When there is a relatively large person sample size, it may be more useful to present the results as they relate to individual persons or subsets of the person sample as they are relevant to the purpose of the analysis.

```
# Calculate average ratings for students:
Person_Avg_Rating <- apply(writing.resp, 1, mean)

# Combine person calibration results in a table:
PC_MFRM_Table4 <- cbind.data.frame(rownames(student.
    locations_PCMFR),
                        Person_Avg_Rating,
                        student.locations_PCMFR$
                        theta,
                        student.locations_PCMFR$se,
                        student.fit$outfitPerson,
                        student.fit$outfitPerson_t,
                        student.fit$infitPerson,
                        student.fit$infitPerson_t)

names(PC_MFRM_Table4) <- c("Student ID", "Average
    Rating", "Student Location","Student SE","Outfit
    MSE","Std. Outfit", "Infit MSE","Std. Infit")
```

```
# Round the numeric values (all columns except the
  first one) to 2 digits:
PC_MFRM_Table4[, -1] <- round(PC_MFRM_Table4[,-1],
  digits = 2)

# Print the first six rows of the table to the
  console
print.data.frame(PC_MFRM_Table4[1:6,], row.names =
  FALSE)
```

```
## Student ID Average Rating Student Location
##          1           2.00             1.88
##          2           2.25             3.04
##          3           1.75             2.66
##          4           2.50            -2.48
##          5           2.75            -0.53
##          6           2.50            -4.05
## Student SE Outfit MSE Std. Outfit Infit MSE
##       0.21       0.97       -0.16       0.97
##       0.21       1.25        1.83       1.29
##       0.21       0.59       -3.26       0.59
##       0.21       1.37        2.27       1.34
##       0.21       0.79       -1.39       0.79
##       0.22       0.80       -1.79       0.82
## Std. Infit
##       -0.13
##        2.04
##       -3.25
##        2.11
##       -1.43
##       -1.67
```

6.3 Notes on Formulations for Many-Facet Rasch Models

In this chapter, we included two examples that demonstrated some basic principles for applying the MFRM using the *TAM* package. Example 1 used a Rating Scale (RS) model formulation with one explanatory person-related facet, and Example 2 used a Partial Credit (PC) formulation with one explanatory item-related facet.

TABLE 6.1
Model Summary Table

Model	TAM Specification
Dichotomous Model	formula <-~item + facet$_i$ + ... + facet$_j$
Rating Scale Model	formula <-~item + facet$_i$ + ... + step
Partial Credit Model	formula <-~item + facet$_i$ + ... + (item:step)
Interaction between facet i and facet j	formula <-~item + facet$_i$ + facet$_j$ + (facet$_i$*facet$_j$) + step

Table 6.1 provides an overview of some popular MFRM formula specifications for the TAM package that may be useful for analyses that differ from those used in this chapter. For example, analysts can specify a dichotomous MFRM by omitting "step" from the formula. In addition, analysts can specify an interaction between facets using the multiplication symbol (*) to create interaction terms.

6.4 Example Results Section

```
# Print Table 6.2:
tab1 <- knitr::kable(
 RS_MFRM_Table1 , booktabs = TRUE,
  caption = 'Model Summary Table'
)
tab1 %>%
  kable_styling(latex_options = "scale_down", full_
    width = FALSE)
```

Table 6.2 presents a summary of the results from the analysis of the style ratings using a Rating Scale model formulation (Andrich, 1978) of the Many-Facet Rasch model (Linacre, 1989).

Specifically, Table 6.2 summarizes the calibration of the students ($N = 372$), subgroups ($N = 2$), and raters ($N = 21$) using average logit-scale calibrations, standard errors, and model-data fit statistics. Student locations represent students' estimated achievement level related to the style of their writing. Higher locations indicate higher achievement. The subgroup facet locations reflect the location of the student language subgroups on the logit scale. Finally, rater locations reflect the severity level of raters when scoring student performances; higher locations indicate more-severe raters. On average, the students were located higher on the logit scale ($M = 0.47$, $SD = 3.16$),

TABLE 6.2

Model Summary Table

Statistic	Students	Subgroups	Raters
Logit Scale Location Mean	0.47	0.00	0.00
Logit Scale Location SD	3.17	0.09	0.52
Standard Error Mean	0.47	0.02	0.09
Standard Error SD	0.12	0.00	0.06
Outfit MSE Mean	0.89	0.95	0.95
Outfit MSE SD	0.35	0.19	0.19
Infit MSE Mean	0.90	0.96	0.96
Infit MSE SD	0.35	0.17	0.17
Std. Outfit Mean	−0.32	−0.55	−0.55
Std. Outfit SD	1.13	1.82	1.82
Std. Infit Mean	−0.30	−0.50	−0.50
Std. Infit SD	1.12	1.85	1.85

compared to raters ($M = 0.00$, $SD = 0.52$), whose locations were centered at zero logits. The average value of the Standard Error (SE) was larger and more variable for students ($M = 0.47$, $SD = 0.12$) compared to raters ($M = 0.09$, $SD = 0.06$). The average values of model-data fit statistics were slightly lower than the expected value of 1.00 for all three facets, indicating that there was slightly less variation in the ratings than expected by the probabilistic model. Additional investigation into item fit and person fit is warranted.

```
# Print Table 6.3:
tab2 <- knitr::kable(
  RS_MFRM_Table2, booktabs = TRUE,
  caption = 'Rater Calibrations'
)
tab2 %>%
  kable_styling(latex_options = "scale_down", full_
    width = FALSE)
```

Table 6.3 includes detailed results for the 21 raters included in the analysis, where raters are ordered by their overall logit-scale location (i.e., rater severity) from high (severe) to low (lenient). For each rater, the average rating is presented, followed by the overall logit-scale location (λ), the location of the rater-specific rating scale category thresholds, and fit statistics specific to the two student subgroups. Because we used a RS formulation of the MFRM, the distance between adjacent rating scale category threshold estimates is the same for all of the raters. Rater 9 was the most severe rater (*Average Rating* = 1.35; $\lambda = 1.12$), and Rater 10 was the most lenient rater (*Average Rating* = 1.82;

TABLE 6.3
Rater Calibrations

Rater ID	Average Rating	Rater Location	Threshold 1	Threshold 2	Threshold 3	Outfit_MSE_Group1	Outfit_MSE_Group2	Infit_MSE_Group1	Infit_MSE_Group2	Std.Outfit_MSE_Group1	Std.Outfit_MSE_Group2	Std.Infit_MSE_Group1	Std.Infit_MSE_Group2
9	1.35	1.12	-2.89	1.35	4.89	0.68	0.89	0.69	0.87	-3.04	-1.04	-3.22	-1.30
19	1.45	0.66	-3.35	0.90	4.44	1.24	1.09	1.15	1.12	1.96	0.87	1.39	1.18
21	1.49	0.50	-3.50	0.74	4.28	0.97	0.79	1.03	0.85	-0.20	-1.86	0.32	-1.50
17	1.51	0.45	-3.56	0.68	4.22	1.14	1.19	1.16	1.17	1.14	1.63	1.48	1.65
16	1.51	0.44	-3.57	0.67	4.21	0.77	0.84	0.78	0.84	-2.14	-1.59	-2.21	-1.75
20	1.51	0.44	-3.57	0.67	4.21	1.01	0.82	1.04	0.85	0.11	-1.79	0.39	-1.56
8	1.55	0.27	-3.74	0.50	4.04	0.75	1.00	0.78	0.98	-2.34	0.03	-2.19	-0.14
14	1.55	0.26	-3.75	0.49	4.03	1.20	1.24	1.19	1.27	1.62	2.10	1.68	2.53
3	1.57	0.17	-3.84	0.40	3.94	0.71	0.97	0.75	1.00	-2.82	-0.27	-2.64	0.03
18	1.59	0.11	-3.90	0.35	3.89	1.05	1.02	1.05	1.05	0.44	0.24	0.48	0.53
4	1.60	0.03	-3.98	0.27	3.81	0.81	0.86	0.83	0.86	-1.75	-1.39	-1.65	-1.48
11	1.66	-0.21	-4.22	0.02	3.56	1.00	1.10	1.01	1.13	0.01	0.88	0.12	1.32
7	1.67	-0.22	-4.23	0.01	3.55	1.10	1.16	1.07	1.07	0.79	1.41	0.70	0.77
1	1.68	-0.28	-4.29	-0.04	3.49	0.54	0.52	0.57	0.55	-4.84	-5.66	-4.91	-5.55
12	1.68	-0.28	-4.29	-0.04	3.49	0.77	0.84	0.78	0.83	-2.14	-1.59	-2.21	-1.79
2	1.69	-0.31	-4.32	-0.08	3.46	0.99	1.16	1.01	1.19	0.00	1.42	0.14	1.87
6	1.69	-0.34	-4.35	-0.10	3.44	0.95	1.06	0.96	1.06	-0.45	0.61	-0.34	0.61
13	1.69	-0.34	-4.35	-0.10	3.44	0.87	0.96	0.88	0.95	-1.13	-0.31	-1.10	-0.46
5	1.80	-0.78	-4.79	-0.54	3.00	0.63	0.99	0.66	1.00	-3.64	-0.07	-3.70	0.00
15	1.81	-0.82	-4.83	-0.59	2.95	0.84	1.00	0.85	0.98	-1.44	0.05	-1.44	-0.15
10	1.82	-0.87	-4.88	-0.63	2.91	1.09	1.24	1.10	1.20	0.82	2.17	0.95	1.98

TABLE 6.4
Student Calibration

Student ID	Average Rating	Student Location	Student SE	Outfit MSE	Std. Outfit	Infit MSE	Std. Infit
69	3.00	7.60	1.47	0.02	−0.81	0.03	−1.02
44	2.95	6.57	0.87	0.54	−0.36	0.71	−0.18
45	2.95	6.57	0.87	0.44	−0.54	0.68	−0.23
12	2.95	6.44	0.87	0.54	−0.36	0.71	−0.18
70	2.95	6.44	0.87	0.76	−0.01	0.79	−0.06
269	2.95	6.44	0.87	0.54	−0.36	0.71	−0.18
77	2.90	5.87	0.69	0.58	−0.58	0.75	−0.33
282	2.86	5.61	0.60	0.83	−0.20	0.93	−0.03
362	2.86	5.61	0.60	1.03	0.22	0.85	−0.24
65	2.86	5.48	0.60	1.13	0.41	0.95	0.02

$\lambda = -0.87$). In general, the raters exhibited similar model-data fit patterns within subgroups.

```
# Print Table 6.3:
tab3 <- knitr::kable(
  head(RS_MFRM_Table3,10), booktabs = TRUE,
  caption = 'Student Calibration'
)
tab3 %>%
  kable_styling(latex_options = "scale_down", full_
    width = FALSE)
```

Table 6.4 includes detailed results for the students who participated in the writing assessment. For each student, the average rating is presented, followed by their logit-scale location estimate (θ), SE, and model-data fit statistics. Students are ordered by their location on the logit scale, from high (high estimated achievement) to low (low estimated achievement). For brevity, Table 6.4 only shows results for 10 students.

```
# Print Table 6.4:
tab4 <- knitr::kable(
  RS_MFRM_Table4, booktabs = TRUE,
  caption = 'Subgroup Calibration'
)
tab4 %>%
  kable_styling(latex_options = "scale_down", full_
    width = FALSE)
```

TABLE 6.5
Subgroup Calibration

Subgroup	Average Rating	Subgroup Location	Subgroup Location SE	Outfit MSE	Std. Outfit	Infit MSE	Std. Infit
2 language2	1.59	0.07	0.02	0.93	−0.24	0.94	−0.21
1 language1	1.63	−0.07	0.02	0.85	−0.41	0.86	−0.41

Table 6.5 shows the calibration of the student subgroup facet. For each subgroup, the average rating among students within the subgroup is presented, followed by the logit-scale location estimate for the subgroup (γ), SE, and model-data fit statistics. Subgroups are ordered by their location on the logit scale, from high (high estimated achievement) to low (low estimated achievement). In this assessment, the difference in logit-scale locations between the two language subgroups was small—indicating comparable estimated achievement levels, on average, for students in each language group. In addition, the fit statistics were comparable between subgroups.

```
# Plot the Wright Map
wrightMap(thetas = cbind(student.locations_RSMFR$
    theta, subgroup.estimates$xsi),
            axis.persons = "Students",
            dim.names = c("Students", "Subgroups"),
            thresholds = rater_thresholds,
            show.thr.lab  = TRUE,
            label.items.rows= 2,
            label.items = rater.estimates$parameter,
            axis.items = "Raters",
            main.title = "Rating Scale Many-Facet Rasch
                Model \nWright Map: Style Ratings",
            cex.main = .6)
```

Rating Scale Many–Facet Rasch Model
Wright Map: Style Ratings

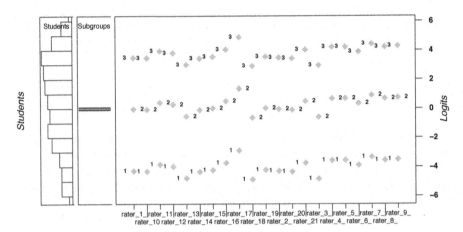

```
##               [,1]          [,2]         [,3]
##    [1,]   -4.288967   -0.04481525   3.493253
##    [2,]   -4.322646   -0.07849438   3.459574
##    [3,]   -3.840553    0.40359889   3.941667
##    [4,]   -3.975059    0.26909229   3.807160
##    [5,]   -4.785429   -0.54127709   2.996791
##    [6,]   -4.345107   -0.10095584   3.437112
##    [7,]   -4.232865    0.01128652   3.549355
##    [8,]   -3.739598    0.50455410   4.042622
##    [9,]   -2.891467    1.35268451   4.890753
##   [10,]   -4.876566   -0.63241444   2.905654
##   [11,]   -4.221649    0.02250299   3.560571
##   [12,]   -4.288967   -0.04481525   3.493253
##   [13,]   -4.345107   -0.10095584   3.437112
##   [14,]   -3.750819    0.49333235   4.031400
##   [15,]   -4.830952   -0.58680019   2.951268
##   [16,]   -3.571092    0.67305934   4.211127
##   [17,]   -3.559845    0.68430692   4.222375
##   [18,]   -3.896607    0.34754470   3.885613
##   [19,]   -3.345721    0.89843079   4.436499
##   [20,]   -3.571092    0.67305934   4.211127
##   [21,]   -3.504491    0.73966026   4.277728
```

Figure 6.1 is a *Wright Map* that illustrates the results from the RS-MFRM analysis of the style ratings. Units on the logit scale are shown on the far-right axis of the plot (labeled *Logits*). The left-most panel of the plot shows a histogram of student locations on the logit scale that represents the latent variable. The second panel from the left shows the distribution of subgroups on the logit scale. There are only two subgroups in our analysis.

The large central panel of the plot shows the rating scale category threshold estimates specific to each rater on the logit scale that represents the latent variable. Light grey diamond shapes show the logit-scale location of the threshold estimates for each rater, as labeled on the x-axis. Thresholds are labeled using an integer that shows the threshold number. In our example, τ_1 is the threshold between rating scale categories $x = 0$ and $x = 1$, τ_2 is the threshold between rating scale categories $x = 1$ and $x = 2$, and τ_3 is the threshold between rating scale categories $x = 2$ and $x = 3$. Because we used a RS model formulation, the distance between adjacent thresholds is the same for all of the raters in the analysis.

The Wright Map suggests that, on average, the students are located higher on the logit scale compared to the average rater threshold locations. In addition, there appears to be a relatively wide spread of student and rater locations on the logit scale, such that the style writing assessment appears to be a useful tool for identifying differences in students' writing achievement related to style as well as differences in rater severity. The subgroup locations are close together, suggesting that there is not much difference in the logit-scale locations between students in either language subgroup.

6.5 Exercise

Please use the *TAM* package to estimate item, threshold, and person locations with the RS-MFRM for the Exercise 6 data. The Exercise 6 data include responses from 350 participants from two subgroups (group 1 and group 2) to a survey with 30 items. Participants used a four-category rating scale ($x = 0, 1, 2, 3$) to respond to each item. The MFRM can be specified using various formulations, including a RS-MFRM and a PC-MFRM. After completing the analysis, try writing a results section similar to the example in this chapter to describe your findings.

7

Basics of Differential Item Functioning

This chapter provides a basic overview of differential item functioning (DIF) along with methods for detecting DIF with Rasch models using R. We will use the *eRm* package (Mair et al., 2021) and the *TAM* package (Robitzsch et al., 2021) for the analyses in this chapter. We have included demonstrations for exploring DIF with dichotomous data and polytomous data using the dichotomous Rasch Model and the Partial Credit Model, respectively.

What is Differential Item Functioning (DIF)?

DIF occurs when examinees who are members of different groups (e.g., demographic subgroups) who have the *same location* on the latent variable have *different probabilities for a response* in a given category (e.g., a correct response or a rating of *Agree*) on an item. In the context of Rasch measurement theory (Rasch, 1960), DIF means that an item has a different location (i.e., a different level of difficulty) between subgroups by more than the standard error. Rasch analyses of DIF typically focus on detecting *uniform DIF*, which occurs when the distance between item response functions (IRFs) is constant between subgroups.

Figure 7.1 illustrates uniform DIF between two groups of examinees for a dichotomous item ($x = 0, 1$). The x-axis shows examinee locations on the latent variable, and the y-axis shows the probability for a correct or positive response ($P(x = 1)$). The dashed line shows the IRF for Group 1, and the dashed line shows IRF for Group 2. This item shows DIF because examinees who have the same location on the latent variable have different probabilities for a correct or positive response.

The interpretation and use of DIF indices varies along with the individual purpose and consequences of each assessment. As a result, it is critical that analysts consider the unique context in which they are evaluating DIF when they interpret the results from DIF analyses. In addition, it is often useful to consider DIF from multiple perspectives, including different statistical techniques for identifying DIF.

It is also important to note that when DIF occurs, it does not always imply a threat to fairness from a psychometric perspective (AERA et al., 2014). Instead, evidence of DIF is a *starting place* for additional research, such as qualitative analyses of item content or the assessment context, to evaluate

Item i

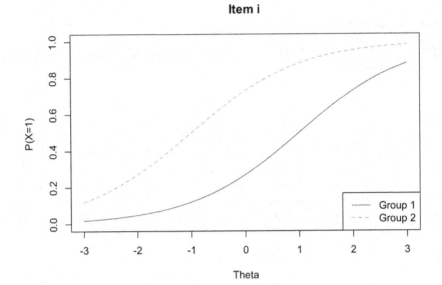

FIGURE 7.1

Uniform DIF between two groups of examinees

whether DIF may reflect a threat to fairness. Some authors (e.g., Myers et al., 2006) refer to this distinction as *statistical DIF*, which occurs when an item is flagged for DIF based on statistical criteria versus *substantive DIF*, which occurs when an item has been flagged for DIF based on statistical criteria *and* substantive review of the item indicates that a construct other than the construct of interest for the assessment can explain differences in item difficulty between subgroups.

In this chapter, we demonstrate two techniques that analysts can use to identify DIF within the Rasch measurement theory framework. Our presentation is not exhaustive and there are many other methods that can be used to supplement those that we demonstrate here. We encourage interested readers to consult the resources listed throughout the chapter as well as the resource list at the end of this chapter to learn more about best practices for identifying DIF within the context of Rasch measurement theory, as well as in other measurement frameworks.

Popular Methods for Identifying DIF with Rasch models

From the perspective of Rasch measurement theory, DIF poses a threat to measurement quality because it implies a lack of invariance. When DIF occurs, the requirements for invariant measurement are not met (see Chapter 1). Accordingly, DIF can be interpreted as not only a potential threat to fairness, but also as a violation of Rasch model requirements (Hagquist and Andrich, 2017a).

Researchers have proposed several methods for identifying DIF within the framework of Rasch measurement theory. Methods that are routinely used in practical applications of Rasch measurement theory can be broadly classified into two approaches: (1) Comparison of group-specific item estimates, and (2) Interaction analyses between item locations and examinee subgroup indicators. In recent methodological work, several researchers have proposed other methods for detecting DIF within the Rasch framework, such as methods based on analysis of variance with model residuals (e.g., Hagquist and Andrich, 2017b) and tree-based methods (Komboz et al., 2018). However, these methods are relatively less common in current applied literature. We encourage interested readers to explore these approaches as well as those that we present here in order to develop a more complete understanding of Rasch-based methods for identifying DIF.

In addition to methods based on Rasch measurement theory, there are numerous other approaches to detecting DIF that are beyond the scope of this book. We encourage readers who are interested in learning more about DIF to consult the resources that we have listed at the end of this chapter.

Comparing Group-Specific Item Estimates

A popular method for identifying DIF within the framework of Rasch measurement theory is to calculate group-specific estimates of item difficulty, and compare those estimates to see if they are meaningfully different. In order to meaningfully compare item difficulty estimates between two groups, it is necessary to adjust (i.e., equate) the item difficulty measures for one group so that they are on the same scale as the item difficulty measures for the second group. The equated differences in item difficulty estimates reflect the distance between group-specific IRFs (Raju, 1988).

Some software programs, including the *eRm* package (Mair et al., 2021), perform this adjustment procedure automatically as part of their DIF functions. In other cases, the analyst needs to perform the adjustment procedure manually before item locations can be compared. For a discussion of the adjustment procedure, please see Luppescu (1995).

After group-specific item difficulty estimates have been calculated and adjusted for the purpose of comparison, it is necessary to consider the degree to which differences in item difficulty estimates are large enough to indicate substantial DIF. Researchers who use Rasch models to evaluate DIF evaluate these differences using several approaches. For example, many researchers calculate the difference between item difficulty estimates as:

$$\delta_{i1} - \delta_{i2} \qquad (7.1)$$

In Equation 7.1, δ_{i1} is the group-specific item difficulty estimate for Group 1, and δ_{i2} is the group-specific item difficulty estimate for Group 2. Researchers sometimes interpret the difference in item difficulty in logits as an effect size, where larger differences indicate more substantial DIF. Wright and Douglas (1975) recommended that researchers use a critical value of 0.5 logits (i.e., the "half-logit rule") to identify DIF that warrants further investigation. This recommendation is based on the typical range of item difficulty for achievement tests (around -2.5 logits to around 2.5 logits), because a difference of 0.5 logits reflects 10% of the item difficulty scale. Differences of this magnitude may impact the accuracy of the measurement procedure between subgroups. Other researchers have proposed smaller critical values, such as Engelhard and Myford (2003), who recommended a difference of 0.3 logits to identify meaningful DIF.

In the Winsteps software manual, Linacre (2020b) proposed guidelines for interpreting differences in item difficulty that reflect the *Delta Index* from ETS (Dorans and Holland, 1993). He noted that differences less than 0.43 logits can be interpreted as negligible DIF (Category A DIF), differences between 0.43 logits and 0.63 logits can be interpreted as slight-to-moderate DIF (Category B DIF), and differences equal to or greater than 0.64 logits as substantial DIF (Category C DIF). These recommendations reflect a conversion between the Delta index that ETS uses to report Mantel-Haenszel DIF statistics (Holland and Thayer, 1988) and logits, where one Delta unit is equal to 0.426 logits.

In some cases, it may also be useful to use a statistical hypothesis test to evaluate whether the difference in item difficulty estimates between groups is larger than what could be expected by chance. Many researchers use the Rasch separate calibration t-test method (Wright and Stone, 1979). This test evaluates the difference between item difficulty estimates calculated from separate calibrations of the same item that have been equated to a common scale. The t-test is defined as:

$$t_i = \frac{\delta_{i1} - \delta_{i1}}{s_p \sqrt{\frac{1}{n_1} + \frac{1}{n_2}}} \tag{7.2}$$

In Equation 7.2, δ_{i1} and δ_{i2} are defined as in Equation 7.1, and s_p is the pooled variance, defined as:

$$s_p = \sqrt{\frac{(n_1 - 1)\, s_1^2 + (n_2 - 1)\, s_2^2}{(n_1 + n_2) - 2}} \tag{7.3}$$

In Equation 7.3, s_1 is the standard error of estimate for δ_{i1}, s_2 is the standard error of estimate for δ_{i2}, n_1 is the number of test-takers in Group 1, and n_2 is the number of test-takers in Group 2 (Smith, 1996b). The resulting t-statistic can be interpreted using a p-value.

When more than two groups are of interest, it is possible to apply the *t*-test method to compare pairs of groups (e.g., Group A compared to Group B; Group B compared to Group C; Group C compared to Group A). As needed, corrections such as Bonferroni adjustments can be used to reduce the chance of Type 1 errors from multiple comparisons.

Interaction Analyses

Another method for identifying DIF is to examine interactions between item difficulty and examinee subgroup identifiers using the Many-Facet Rasch Model (MFRM); (Linacre, 1989 see Chapter 6). Including interaction terms in MFRMs allows researchers to test the null hypothesis that the calibrations of elements within one facet (e.g., items) are invariant across levels of another facet (e.g., subgroups). For the purpose of identifying DIF, researchers can examine the magnitude of the interaction effect between individual items and relevant examinee subgroups, where larger interaction terms indicate larger differences in item difficulty parameters for the subgroup of interest.

For example, the following formulation of the MFRM could be used to evaluate DIF:

$$\ln\left(\frac{P_{nig(x=k)}}{P_{nig(x=k-1)}}\right) = (\theta_n - \delta_i - \gamma_g - \tau_k) - \delta_i\gamma_g \qquad (7.4)$$

In Equation 7.4, θ_n is the location of Examinee n, δ_i is the location of item i, γ_g is the location of examinee subgroup g, and τ_k is the rating scale category threshold between category k and category $k - 1$. The model also includes the interaction between item locations and subgroup locations: $\delta_i*\gamma_g$. The interaction term tests the null hypothesis that item locations are invariant across examinee subgroups, against the alternative hypothesis that item locations are different across examinee subgroups, which would be evidence of DIF.

7.1 Detecting Differential Item Functioning in R for Dichotomous Items

In this section, we provide a step-by-step demonstration of the two approaches to DIF analyses for dichotomous item response data. We encourage readers to use the example data set that is provided in the online supplement to conduct the analysis along with us.

7.1.1   Example Data: Transitive Reasoning

The example data for this section is the Transitive Reasoning Data that
we used in Chapter 2. The data include a group of 425 children's scored
responses to assessment items designed to measure their ability to reason
about the relationships among physical objects. Along with student responses,
the Transitive Reasoning data file includes students' grade levels, ranging from
Grade 2 to Grade 6. We will use students' grade level as the basis for our DIF
analyses. Please see Chapter 2 for additional details about these data.

Prepare for the Analyses

We will use the *Extended Rasch Modeling* or *eRm* package (Mair et al., 2021)
and the *Test Analysis Modules* or *TAM* package (Robitzsch et al., 2021) to
demonstrate DIF analyses in this chapter.

```
# install.packages("eRm")
library("eRm")


# install.packages("eRm")
library("TAM")
```

Now that we have installed and loaded the packages to our R session, we are
ready to import the data. We will use the function `read.csv()` to import
the comma-separated values (.csv) file that contains the Transitive Reasoning
assessment data. We encourage readers to use their preferred method for
importing data files into R or R Studio.

Please note that if you use `read.csv()` to import the data, you will first need
to specify the file path to the location at which the data file is stored on your
computer or set your working directory to the folder in which you have saved
the data.

We will import the data using `read.csv()` and store it in an object called
`transreas`.

```
transreas <- read.csv("transreas.csv")
```

Next, we need to identify the variable that contains examinees' subgroup
membership identifiers for our DIF analysis. In our example analysis, we will
conduct DIF analyses related to students' grade level. The Transitive Reasoning
assessment data includes five grade levels (Grade 2 through Grade 6). For
the sake of a simple illustration, we will re-code these grade levels into two
groups: (1) Lower Elementary students (Grade 2 through Grade 4) and Upper
Elementary students (Grade 5 and Grade 6).

The following code creates a vector object that includes students' membership
in the Lower Elementary or Upper Elementary groups. Then, we use the
`table()` function to generate a frequency table for this variable.

```
transreas.grade.group <- ifelse(transreas$Grade <= 4,
    1, 2)
```

```
table(transreas.grade.group)
## transreas.grade.group
##   1   2
## 253 172
```

DIF Method 1: Compare Item Locations between Subgroups

First, we will conduct a DIF analysis by comparing the difficulty estimate of each Transitive Reasoning item between our two student subgroups. We will use the dichotomous Rasch model (see Chapter 2) to estimate item difficulty locations.

The *eRm* package includes a function that allows researchers to calculate group-specific item difficulty estimates *after* the item responses have been analyzed for the complete sample. Accordingly, we will begin our analysis by estimating item and person locations for the complete sample of students who completed the Transitive Reasoning assessment using the Rasch model.

To get started with the DIF analysis in *eRm*, we need to isolate the response matrix from our data and then apply the dichotomous Rasch model to the responses. The following code completes these steps. Please refer to Chapter 2 for a detailed description of procedures for applying and interpreting the results from the dichotomous Rasch model.

```
## Isolate the response matrix:
transreas.responses <- subset(transreas, select = -c(
    Student, Grade))
```

```
## Apply the model to the data
transreas.RM <- RM(transreas.responses)
```

Next, we will calculate subgroup-specific item difficulty values using the Waldtest() function from *eRm*. We will specify the transreas.grade.group object as the grouping variable for our analysis.

```
subgroup_diffs <- Waldtest(transreas.RM, splitcr =
    transreas.grade.group)
```

This analysis saves numerous details to the object called subgroup_diffs. Let's create new objects with the group-specific item difficulties:

```
subgroup_1_diffs <- subgroup_diffs$betapar1
subgroup_2_diffs <- subgroup_diffs$betapar2
```

We can examine the differences between the subgroup-specific item difficulties
by subtracting the two sets of values.

```
subgroup_1_diffs - subgroup_2_diffs
```

```
## beta task_01 beta task_02 beta task_03
##     0.3648935    0.5457623    0.4623913
## beta task_04 beta task_05 beta task_06
##     0.2437839   -0.2048691   -0.5690487
## beta task_07 beta task_08 beta task_09
##    -0.3396100   -0.3781807    0.1190287
## beta task_10
##    -0.2441513
```

We can see that there are some items that were easier for Lower Elementary
students (positive differences) and other items that were easier for Upper
Elementary students (negative differences). In addition, the difference in logit-
scale locations for two items (Task 2 and Task 6) exceed the "half-logit" rule,
suggesting potentially meaningful DIF. The difference for Task 3 approximates
this value as well.

To better understand the significance of differences in item difficulty between
subgroups, we will examine the results from the statistical hypothesis test for
the item difficulty comparisons. In the *eRm* package, item comparisons are
reported using z-statistics instead of the Wright and Stone *t*-test statistics
that we discussed earlier. With large sample sizes, t statistics approximate z
statistics. Let's create an object called comparisons in which we store these
results.

```
comparisons <- as.data.frame(subgroup_diffs$coef.
   table)
```

For reporting purposes, it is often helpful to construct a table that includes
subgroup-specific item difficulty values along with the results from the statistical hypothesis test. The following code saves these results in an object called
comparison.results, and then prints the object to the console.

```
comparison.results <- cbind.data.frame(subgroup_1_
   diffs, subgroup_diffs$se.beta1,
                            subgroup_2_
                            diffs,
                            subgroup_
                            diffs$se.
                            beta2,
                            comparisons)
# Name the columns of the results
names(comparison.results) <- c("Subgroup_1_Difficulty
   ", "Subgroup_1_SE",
```

```
                              "Subgroup_2_Difficulty
                            ", "Subgroup_2_SE",
                            "Z", "p_value")
comparison.results
```

```
##                 Subgroup_1_Difficulty  Subgroup_1_SE
## beta task_01             1.35779188      0.2571262
## beta task_02            -0.06336358      0.1702630
## beta task_03             0.57207888      0.1997716
## beta task_04            -0.36534197      0.1607645
## beta task_05            -0.39081998      0.1600840
## beta task_06             2.02881377      0.3290900
## beta task_07            -0.09273638      0.1692194
## beta task_08             1.79979029      0.3017318
## beta task_09            -2.89136085      0.1697609
## beta task_10            -1.95485207      0.1483956
##                 Subgroup_2_Difficulty  Subgroup_2_SE
## beta task_01             0.9928984       0.3383101
## beta task_02            -0.6091259       0.2169509
## beta task_03             0.1096876       0.2562673
## beta task_04            -0.6091259       0.2169508
## beta task_05            -0.1859509       0.2375660
## beta task_06             2.5978625       0.6541650
## beta task_07             0.2468736       0.2662718
## beta task_08             2.1779710       0.5424163
## beta task_09            -3.0103896       0.2095098
## beta task_10            -1.7107008       0.1922855
##                          Z        p_value
## beta task_01     0.8587094    0.39050087
## beta task_02     1.9789435    0.04782237
## beta task_03     1.4230349    0.15472603
## beta task_04     0.9028236    0.36661953
## beta task_05    -0.7151527    0.47451472
## beta task_06    -0.7770931    0.43710383
## beta task_07    -1.0764419    0.28172969
## beta task_08    -0.6092896    0.54233252
## beta task_09     0.4414133    0.65891380
## beta task_10    -1.0051980    0.31480152
```

From these results, we see that there is one item with a statistically significant difference between subgroups based on $p < 0.05$: Task 2.

As an additional interpretation aid, we will construct a scatterplot that shows the alignment between item difficulty estimates for the two subgroups in our analysis. Because the procedure that we used to calculate item difficulties specific to each subgroup has already adjusted the difficulties to be on the

same scale, we don't need to apply any transformation. The following code
will create a scatterplot with 95% confidence bands to highlight items that are
significantly different between the subgroups.

```
## Calculate values for constructing the confidence
   bands:

mean.1.2 <- ((subgroup_1_diffs - mean(subgroup_1_
   diffs))/2*sd(subgroup_1_diffs) +
                (subgroup_2_diffs - mean(subgroup_2_
                   diffs))/2*sd(subgroup_2_diffs))

joint.se <- sqrt((subgroup_diffs$se.beta1^2/sd(
   subgroup_1_diffs)) +
                  (subgroup_diffs$se.beta2^2/sd(
                     subgroup_2_diffs)))

upper.group.1 <- mean(subgroup_1_diffs) + ((mean.1.2
   - joint.se )*sd(subgroup_1_diffs))
upper.group.2 <- mean(subgroup_2_diffs) + ((mean.1.2
   + joint.se )*sd(subgroup_2_diffs))

lower.group.1 <- mean(subgroup_1_diffs) + ((mean.1.2
   + joint.se )*sd(subgroup_1_diffs))
lower.group.2 <- mean(subgroup_2_diffs) + ((mean.1.2
   - joint.se )*sd(subgroup_2_diffs))

upper <- cbind.data.frame(upper.group.1, upper.group
   .2)
upper <- upper[order(upper$upper.group.1, decreasing
   = FALSE),]

lower <- cbind.data.frame(lower.group.1, lower.group
   .2)
lower <- lower[order(lower$lower.group.1, decreasing
   = FALSE),]

## Construct the scatterplot:

plot(subgroup_1_diffs, subgroup_2_diffs, xlim = c(-4,
   4), ylim = c(-4, 4),
     xlab = "Lower Elementary", ylab = "Upper
        Elementary", main = "Lower Elementary
        Measures plotted against Upper Elementary
        Measures", cex.main = .8)
abline(a = 0, b = 1, col = "purple")
```

```
par(new = T)

lines(upper$upper.group.1, upper$upper.group.2, lty =
    2, col = "red")

lines(lower$lower.group.1, lower$lower.group.2, lty =
    2, col = "red")

legend("bottomright", c("Item Location", "Identity
    Line", "95% Confidence Band"),
        pch = c(1, NA, NA), lty = c(NA, 1, 2), col = c
        ("black", "purple", "red"), cex = .8)
```

Lower Elementary Measures plotted against Upper Elementary Measures

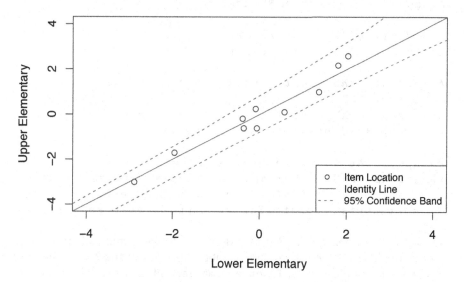

In this scatterplot, item difficulty estimates for the Lower Elementary group are plotted along the x-axis, and item difficulty estimates for the Upper Elementary group are plotted along the y-axis. Dashed lines show a 95% confidence interval beyond which the differences in item difficulty estimates are statistically significant. Items that are located above the identity line have higher locations (i.e., are more difficult) for students in the Upper Elementary group, and items that are located below the identity line are more difficult for students in the Lower Elementary group.

DIF Method 2: Interaction Analysis

Next, we will conduct a DIF analysis with the Transitive Reasoning data using the interaction analysis approach based on the MFRM (see Chapter 6). To do this, we will use the *TAM* package (Robitzsch et al., 2021) to specify a MFRM that includes an interaction term between item difficulty and student subgroup membership.

To get started with the interaction analysis with the MFRM, we need to isolate the response matrix from our data and then specify the model. The following code completes these steps. Please refer to Chapter 6 for a detailed description of procedures for applying and interpreting the results from the MFRM.

```
## Specify the facets in the model (grade-level
    subgroup) as a data.frame object
facets <- as.data.frame(transreas.grade.group)
facets <- facets[,"transreas.grade.group", drop=FALSE
    ]

## Identify the object of measurement (students)
students <- transreas$Student

## Identify the response matrix:
ratings <- transreas.responses

## Specify the model equation with the interaction
    term:
transreas_MFRM_equation <- ~ item + transreas.grade.
    group + (item*transreas.grade.group)

## Apply the model to the responses:
transreas_MFRM <- tam.mml.mfr(resp = ratings, facets
    = facets, formulaA = transreas_MFRM_equation, pid
    = students, constraint = "items", verbose = FALSE)
```

The estimates from the MFRM include values for each item, subgroup, and the interaction between each item and subgroup. We will store all of the facet estimates in an object called `facet.estimates`, and then store the results for each facet in separate objects for easier manipulation.

```
facet.estimates <- transreas_MFRM$xsi.facets

item.estimates <- subset(facet.estimates, facet.
    estimates$facet == "item")

subgroup.estimates <- subset(facet.estimates, facet.
    estimates$facet == "transreas.grade.group")

interaction.estimates <- subset(facet.estimates,
    facet.estimates$facet == "item:transreas.grade.
    group")
```

Our main focus is on the interaction results. We can examine the relative magnitude of each interaction term by plotting the values in a simple scatterplot. The first ten values in the `interaction.estimates` object show the interaction between each item and group 1 (Lower Elementary), and the second ten values show the interaction between each item and group 2 (Upper Elementary). We will construct separate plots for each subgroup.

```
plot(interaction.estimates$xsi[1:10], main = "
    Interaction Effects for Lower Elementary Students"
    , ylab = "Interaction Estimate in Logits", xlab =
    "Item Number", ylim = c(-0.5, 0.5))
```

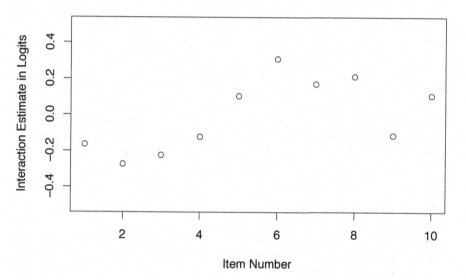

Interaction Effects for Lower Elementary Students

```
plot(interaction.estimates$xsi[11:20], main = "
    Interaction Effects for Upper Elementary Students"
    , ylab = "Interaction Estimate in Logits", xlab =
    "Item Number", ylim = c(-0.5, 0.5))
```

Interaction Effects for Upper Elementary Students

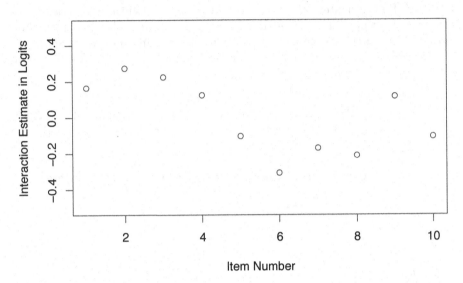

The interaction estimates are reported on the same logit scale as the item and person estimates. Values equal to zero indicate no interaction between item difficulty and subgroup membership. Positive values indicate that an item was relatively easier for a given subgroup compared to its overall location estimate. Negative values indicate that an item was relatively more difficult for a given subgroup compared to its overall location estimate.

We can examine the results and see that the largest interaction terms were observed for Task 2 and Task 6. Task 2 was more difficult for Lower Elementary students compared to Upper Elementary students. The opposite pattern was true for Task 6, which was easier for Lower Elementary Students compared to Upper Elementary students.

7.2   Detecting Differential Item Functioning in R for Polytomous Items

In this section, we demonstrate some basic procedures for detecting DIF with item responses that are scored in more than two categories (i.e., polytomous items). Specifically, we will illustrate DIF analyses based on the Partial Credit Model (Masters, 1982), see Chapter 5. To do this, we will use the style ratings data that we used in the first illustration in Chapter 6. Most of the procedures for the analyses in this section mirror those from the illustrations with the dichotomous Rasch model earlier in this chapter. As a result, we provide relatively less detail in this section except where information is needed to highlight important differences.

7.2.1 Example Data: Style Ratings

The example data for this section are 21 rater judgments of the style of 372 students' written compositions that were collected during a middle-grades writing assessment in the United States. We analyzed these data in Chapter 6. The minimum rating from each rater was $x = 0$, and the maximum rating was $x = 3$. The example data include two demographic subgroups related to students' language background: Subgroup 1 (*language* $= 1$) indicates that students' best language is a language other than English, and subgroup 2 (*language* $= 2$) indicates that students' best language is English. We will use students' language subgroup membership as the basis for our DIF analyses.

For the sake of illustration, we will treat the raters as "pseudo-items" in our analysis. DIF indicators will highlight individual raters who exhibit systematic differences in severity between student subgroups, after controlling for students' location on the latent variable.

Prepare for the Analyses

As we did in the first example, we will use two packages for our polytomous DIF analyses: *eRm* (Mair et al., 2021) and *TAM* (Robitzsch et al., 2021). Please ensure that these packages have been installed and loaded into the R environment (R Core Team, 2016) before proceeding with the analyses.

Next, we will import the style ratings and prepare for the analysis. We will isolate the response matrix and identify the subgroup variable (language).

```
style <- read.csv("style_ratings.csv")

ratings <- subset(style, select = -c(student,
    language))
```

DIF Method 1: Compare Item Locations between Subgroups

It is possible to use the same approach to compare item difficulty between subgroups with polytomous data as with dichotomous data. The major difference is that, with polytomous data, item difficulty estimates include rating scale category thresholds. In this illustration, we will demonstrate how to compare the difficulty of rating scale category thresholds between subgroups, and how to compare overall item difficulty estimates.

For the sake of illustration, we use the Partial Credit Model (PCM) (Masters, 1982) (see Chapter 5) for our DIF analyses. However, similar analyses can also be performed using the Rating Scale Model (RSM) (Andrich, 1978) (see Chapter 4).

To get started with the DIF analysis, we need to apply the PCM to the style ratings.

```
PC_model <- PCM(ratings)
```

The estimation procedure for polytomous data results in item-specific estimates of rating scale category thresholds. If you want to compare the threshold locations between subgroups, you can use: `subgroup_diffs <- Waldtest( PC_model, splitcr = subgroups)` to generate the DIF results specific to item-category thresholds, and then proceed as in the first illustration in this chapter. However, if you want to make comparisons at the overall item level, you will need to calculate item difficulty estimates and standard errors for the overall item. In our example, we are treating raters as "psedo-items", so the overall item estimates will actually be rater severity estimates.

First, we identify subgroup classification variable that we will use for our DIF analysis. In our example, this is the student language variable.

```
subgroups <- style$language
```

Next, we will calculate subgroup-specific item difficulty (rater severity) estimates. We will store these values in objects called `group1_item.diffs. overall` and `group2_item.diffs.overall`.

```
group1_item.diffs.overall <- NULL
group2_item.diffs.overall <- NULL

responses.g <- cbind.data.frame(subgroups, ratings)
responses.g1 <- subset(responses.g, responses.g$
   subgroups == 1)
responses.g2 <- subset(responses.g, responses.g$
   subgroups == 2)

subgroup_diffs <- Waldtest(PC_model, splitcr =
   subgroups)

for(item.number in 1:ncol(ratings)){
  n.thresholds.g1 <-  length(table(responses.g1[,
    item.number+1]))-1

  group1_item.diffs.overall[item.number] <- mean(
    subgroup_diffs$betapar1[(((item.number*(n.
    thresholds.g1))-(n.thresholds.g1-1)):(item.
    number*(n.thresholds.g1))])*-1
```

```
n.thresholds.g2 <-  length(table(responses.g2[, item.
    number+1]))-1

group2_item.diffs.overall[item.number] <- mean(
    subgroup_diffs$betapar2[(((item.number*(n.
    thresholds.g2))-(n.thresholds.g2-1)):(item.number*
    (n.thresholds.g2))])*-1
}
```

We can view the group-specific item difficulty (rater severity) estimates by printing them to the console.

```
group1_item.diffs.overall
```

```
##   [1]  -0.4879483  -1.6685347   0.4046874  -0.2173740
##   [5]  -1.0037163  -1.0198568  -0.6036101   0.9691077
##   [9]   2.7159392  -0.8263795  -0.6954105  -0.8690388
##  [13]  -1.0935391   0.5013196  -1.9835298   0.6158012
##  [17]   0.4703671   0.5245534   1.8621749   1.2878600
##  [21]   1.1171273
```

```
group2_item.diffs.overall
```

```
##   [1]  -0.47182824  -1.09475957   0.02793690
##   [4]   0.26851229  -1.86361946  -0.54677268
##   [7]  -0.52376081   0.57629338   2.55640283
##  [10]  -1.98991035  -0.01255957  -0.24059455
##  [13]  -1.05035379   0.97971490  -1.75389400
##  [16]   0.06969827   0.78191084   0.34034134
##  [19]   1.16799459   1.14577258   1.63347510
```

To compare these values, we need to calculate overall standard errors for each item (rater). In the following code, we save these values in objects called group1_item.se.overall and group2_item.se.overall.

```
group1_item.se.overall <- NULL
group2_item.se.overall <- NULL

responses.g <- cbind.data.frame(subgroups, ratings)
responses.g1 <- subset(responses.g, responses.g$
    subgroups == 1)
responses.g2 <- subset(responses.g, responses.g$
    subgroups == 2)

subgroup_diffs <- Waldtest(PC_model, splitcr =
    subgroups)
```

```
for(item.number in 1:ncol(ratings)){

  n.thresholds.g1 <-  length(table(responses.g1[,
    item.number+1]))-1

  group1_item.se.overall[item.number] <- mean(subgroup_
    diffs$se.beta1[((item.number*(n.thresholds.g1))-(n
    .thresholds.g1-1)):(item.number*(n.thresholds.g1))
    ])

  n.thresholds.g2 <-  length(table(responses.g2[, item.
    number+1]))-1

  group2_item.se.overall[item.number] <- mean(subgroup_
    diffs$se.beta2[((item.number*(n.thresholds.g2))-(n
    .thresholds.g2-1)):(item.number*(n.thresholds.g2))
    ])
}
```

Then, we will calculate test statistics (Z) for the differences in overall item difficulties. We will print the results to the console.

```
z <- (group1_item.diffs.overall - group2_item.diffs.
    overall)/
    sqrt(group1_item.se.overall^2 + group2_item.se.
    overall^2)
z
```

```
##   [1]  -0.02664874  -0.85932508   0.62997905
##   [4]  -0.80736274   1.36218140  -0.75197550
##   [7]  -0.12854486   0.67861431   0.28401849
##  [10]   1.89639423  -1.13050332  -1.03552438
##  [13]  -0.06716818  -0.82354357  -0.35040548
##  [16]   0.88195315  -0.51329786   0.31338943
##  [19]   1.21102675   0.24832588  -0.90554145
```

Finally, we will construct a scatterplot of item (rater) measures with 95% confidence interval bands for the two subgroups.

```
## First, calculate values for constructing the
    confidence bands:

mean.1.2 <- ((group1_item.diffs.overall - mean(group1
    _item.diffs.overall))/2*sd(group1_item.diffs.
    overall) +
```

```
           (group2_item.diffs.overall - mean(
              group2_item.diffs.overall))/2*sd(
              group2_item.diffs.overall))

joint.se <- sqrt((group1_item.se.overall^2/sd(group1_
   item.diffs.overall)) + (group2_item.se.overall^2/
   sd(group2_item.diffs.overall)))


upper.group.1 <- mean(group1_item.diffs.overall) + ((
   mean.1.2 - joint.se )*sd(group1_item.diffs.overall
   ))

upper.group.2 <- mean(group2_item.diffs.overall) + ((
   mean.1.2 + joint.se )*sd(group2_item.diffs.overall
   ))

lower.group.1 <- mean(group1_item.diffs.overall) + ((
   mean.1.2 + joint.se )*sd(group1_item.diffs.overall
   ))

lower.group.2 <- mean(group2_item.diffs.overall) + ((
   mean.1.2 - joint.se )*sd(group1_item.diffs.overall
   ))

upper <- cbind.data.frame(upper.group.1, upper.group
   .2)
upper <- upper[order(upper$upper.group.1, decreasing
   = FALSE),]

lower <- cbind.data.frame(lower.group.1, lower.group
   .2)
lower <- lower[order(lower$lower.group.1, decreasing
   = FALSE),]

## make the scatterplot:

plot(group1_item.diffs.overall, group2_item.diffs.
   overall, xlim = c(-4, 4), ylim = c(-4, 4),
      xlab = "Language Other than English", ylab = "
         English", main = "Language-other-than-English
         Measures \n plotted against English Measures
         ")
abline(a = 0, b = 1, col = "purple")
```

```
par(new = T)

lines(upper$upper.group.1, upper$upper.group.2, lty =
    2, col = "red")

lines(lower$lower.group.1, lower$lower.group.2, lty =
    2, col = "red")

legend("bottomright", c("Item Location", "Identity
    Line", "95% Confidence Band"), pch = c(1, NA, NA),
    lty = c(NA, 1, 2), col = c("black", "purple", "
    red"))
```

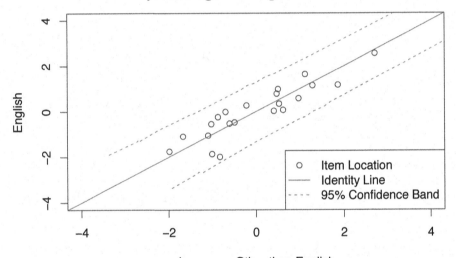

Language–other–than–English Measures plotted against English Measures

DIF Method 2: Interaction Analysis

Next, we will conduct a DIF analysis with the style ratings data using the interaction analysis approach based on the MFRM (see Chapter 6). We will use the *TAM* package (Robitzsch et al., 2021) to specify a MFRM that includes an interaction term between rater severity and student subgroup membership.

The following code applies the MFRM to the style rating data. Please refer to Chapter 6 for a detailed description of procedures for applying and interpreting the results from the MFRM.

```
## Specify the facets in the model (language subgroup
    ) as a data.frame object
facets <- style[,"language", drop=FALSE]
```

```
## Identify the object of measurement (students)
students <- style$student

## Identify the response matrix:
ratings <- style[, -c(1:2)]

## Specify the model equation with the interaction
    term:
style_PC_MFRM_equation <- ~ item + language + item:
    step + (item*language)

## Apply the model to the responses:
style_MFRM <- tam.mml.mfr(resp = ratings, facets =
    facets, formulaA = style_PC_MFRM_equation, pid =
    students, constraint = "items", verbose = FALSE)
```

Next, we store all of the facet estimates in an object called `facet.estimates`, and then store the results for each facet in separate objects for easier manipulation. Note that the `item.estimates` object actually includes rater severity estimates.

```
facet.estimates <- style_MFRM$xsi.facets

item.estimates <- subset(facet.estimates, facet.
    estimates$facet == "item")

subgroup.estimates <- subset(facet.estimates, facet.
    estimates$facet == "language")

threshold.estimates <- subset(facet.estimates, facet.
    estimates$facet == "item:step")

interaction.estimates <- subset(facet.estimates,
    facet.estimates$facet == "item:language")
```

We can examine the relative magnitude of each interaction term by plotting the values in a simple scatterplot. The first 21 values in the `interaction .estimates` object show the interaction between each rater and group 1 (language other than English), and the second 21 values show the interaction between each rater and group 2 (English). We will construct separate plots for each subgroup.

```
plot(interaction.estimates$xsi[1:21], main = "
    Interaction Effects for Language Other than
    English", ylab = "Interaction Estimate in Logits",
```

```
xlab = "Item (Rater) Number", ylim = c(-0.5, 0.5)
)
```

Interaction Effects for Language Other than English

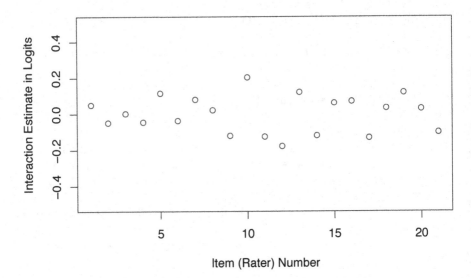

plot(interaction.estimates$xsi[22:42], main = "
    Interaction Effects for English", ylab = "
    Interaction Estimate in Logits", xlab = "Item (
    Rater) Number", ylim = c(-0.5, 0.5))

Interaction Effects for English

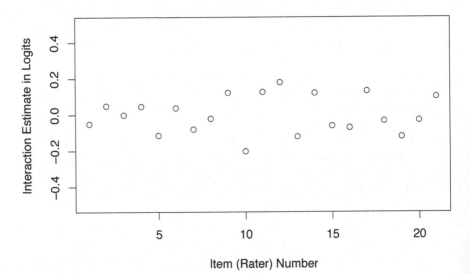

7.3 Exercise

There are two exercises for this chapter. The first exercise uses dichotomous data, and the second exercise uses polytomous data.

The Exercise 7A data include responses from 350 students in two subgroups (group 1 and group 2) who responded to a multiple-choice assessment with 20 items. The items were scored such that $x = 1$ indicates a correct response and $x = 2$ indicates an incorrect response. Try analyzing the data for evidence of DIF.

The Exercise 7B data include responses from 400 participants from two subgroups (group 1 and group 2) to a survey with 25 items. Participants used a four-category rating scale ($x = 0, 1, 2, 3$) to respond to each item. Try analyzing the data for evidence of DIF.

Bibliography

AERA, APA, and NCME (2014). *Standards for Educational and Psychological Testing*. American Educational Research Association, United States.

Alemdar, M., Lingle, J. A., Moore, R., and Wind, S. A. (2017). Developing an engineering design process assessment using think-aloud interviews. *International Journal of Engineering Education*, 33:441–452.

Andrich, D. (1978). A rating formulation for ordered response categories. *Psychometrika*, 43(4):561–573.

Andrich, D. (1982). An index of person separation in latent trait theory, the traditional KR-20 index, and the Guttman scale response pattern. *Education Research and Perspectives*, 9(1).

Andrich, D. (2015). The problem with the step metaphor for polytomous models for ordinal assessments. *Educational Measurement: Issues and Practice*, 34(2):8–14.

Andrich, D. and Marais, I. (2019). *A Course in Rasch Measurement Theory: Measuring in the Educational, Social and Health Sciences*. Springer, Singapore.

Birnbaum, A. (1968). *Some Latent Trait Models and Their Use in Inferring an Examinee's Ability*, pages 397–479. Statistical Theories of Mental Test Scores. Addison-Wesley, Menlo Park.

Bond, T., Yan, Z., and Heene, M. (2019). *Applying the Rasch Model: Fundamental Measurement in the Human Sciences*. Routledge, New York, 4th edition.

Bonifay, W. (2020). *Multidimensional Item Response Theory*. SAGE Publications, 1st edition.

Borsboom, D. and Mellenbergh, G. J. (2004). Why psychometrics is not pathological: A comment on Michell. *Theory and Psychology*, 14:105–120.

Briggs, D. and Wilson, M. (2003). An introduction to multidimensional measurement using Rasch models. *Journal of Applied Measurement*, 4(1).

Chou, Y.-T. and Wang, W.-C. (2010). Checking dimensionality in item response models with principal component analysis on standardized residuals. *Educational and Psychological Measurement*, 70(5):717–731.

Cronbach, L. J. (1951). Coefficient alpha and the internal structure of tests. *Psychometrika*, 16.

Dorans, N. J. and Holland, P. W. (1993). *DIF Detection and Description: Mantel-Haenszel and Standardization*, pages 35–66. Differential item functioning. Lawrence Erlbaum Associates, Inc.

Embretson, S. E. and Reise, S. P. (2000). *Item Response Theory for Psychologists*. Lawrence Erlbaum Associates.

Engelhard, G. and Wang, J. (2020). *Rasch Models for Solving Measurement Problems: Invariant Measurement in the Social Sciences*, volume 187. SAGE, 4th edition.

Engelhard, G. and Wind, S. A. (2018). *Invariant Measurement with Raters and Rating Scales: Rasch Models for Rater-Mediated Assessments*. Taylor and Francis.

Engelhard, G. J. and Myford, C. M. (2003). *Monitoring Faculty Consultant Performance in the Advanced Placement English Literature and Composition Program with a Many-Faceted Rasch Model*. College Entrance Examination Board.

Hagquist, C. and Andrich, D. (2017a). Recent advances in analysis of differential item functioning in health research using the Rasch model. *Health and Quality of Life Outcomes*, 15(181):1–8.

Hagquist, C. and Andrich, D. (2017b). Recent advances in analysis of differential item functioning in health research using the Rasch model. *Health and Quality of Life Outcomes*, 15(181).

Holland, P. W. and Thayer, D. (1988). *Differential Item Performance and the Mantel-Haenszel Procedure*, pages 129–145. Test validity. Lawrence Erlbaum Associates, Inc.

Irribarra, D. T. and Freund, R. (2014). *Wright Map: IRT Item-Person Map with ConQuest Integration*.

Komboz, B., Strobl, C., and Zeileis, A. (2018). Tree-based global model tests for polytomous Rasch models. *Educational and Psychological Measurement*, 78(1):128–166.

Linacre, J. M. (1989). *Many-Facet Rasch Measurement*. MESA Press.

Linacre, J. M. (2002). Optimizing rating scale category effectiveness. *Journal of Applied Measurement*, 3(1):85–106.

Linacre, J. M. (2003). Data variance: Explained, modeled and empirical. *Rasch Measurement Transactions*, 17(3):942–943.

Linacre, J. M. (2016). Principal-components analysis of residuals. https://www.winsteps.com/winman/principalcomponents.htm.

Linacre, J. M. (2020a). *Facets Computer Program for Many-Facet Rasch Measurement.* Beaverton, Oregon.

Linacre, J. M. (2020b). *A User's Guide to WINSTEPS MINISTEP Rasch-Model Computer Programs.*

Luppescu, S. (1995). Comparing measures: Scatterplots. *Rasch Measurement Transactions,* 9(1):410.

Mair, P., Hatzinger, R., and Maier, M. J. (2021). *eRm: Extended Rasch Modeling.* 1.0-2.

Masters, G. (1982). A Rasch model for partial credit scoring. *Psychometrika,* 47(2):149–174.

Maul, A. (2020). Ask an expert: Rasch and dimensionality. *Rasch Measurement Transactions,* 33(1):1753–1754.

Mellenbergh, G. J. (1995). Conceptual notes on models for discrete polytomous item responses. *Applied Psychological Measurement,* 19(1):91–100.

Michael.Linacre, J. (1998). Structure in Rasch residuals: Why principal components analysis (pca)? *Rasch Measurement Transactions,* 12(2):636.

Michell, J. (1999). *Measurement in Psychology: A Critical History of a Methodological Concept.* Cambridge University Press, Cambridge, England.

Myers, N. D., Wolfe, E. W., Feltz, D. L., and Penfield, R. D. (2006). Identifying differential item functioning of rating scale items with the Rasch model: An introduction and an application. *Measurement in Physical Education and Exercise Science,* 10(4).

Nam, S., Kyung, E., Lee, S. M., Lee, S. H., and Seol, H. (2011). A psychometric evaluation of the career decision self-efficacy scale with Korean students: A Rasch model Spproach. *Journal of Career Development,* 38(2):147–166.

R Core Team (2016). *R: A Language and Environment for Statistical Computing.* R Foundation for Statistical Computing, Vienna, Austria.

R Core Team (2021). *R: A Language and Environment for Statistical Computing.* R Foundation for Statistical Computing, Vienna, Austria.

Raju, N. S. (1988). The area between two item characteristic curves. *Psychometrika,* 53(4):495–502.

Rasch, G. (1960). *Probabilistic Models for Some Intelligence and Achievement tests.* University of Chicago Press, Vienna, Austria, expanded edition, 1980 edition.

Rasch, G. (1961). Proceedings of the IV Berkeley symposium on mathematical statistics and probability. In *On General Laws and the Meaning of Measurement in Psychology,* volume 4.

Rasch, G. (1977). On specific objectivity: An attempt at formalizing the request for generality and validity of scientific statements. *Danish Yearbook of Psychology*, 14(58–94).

Rasch, G. (2005). Critical eigenvalue sizes in standardized residual principal components analysis. *Rasch Measurement Transactions*, 19:1012.

Reckase, M. D. (1979). Unifactor latent trait models applied to multifactor tests: Results and implications. *Journal of Educational and Behavioral Statistics*, 4(3):207–230.

Reckase, M. D. (2009). *Multidimensional Item Response Theory*. Springer, New York, NY. ISBN 978-0-387-89975-6.

Revelle, W. (2021). *psych: Procedures for Psychological, Psychometric, and Personality Research*. Northwestern University, Evanston, Illinois. R package version 2.1.6.

Robitzsch, A., Kiefer, T., and Wu, M. (2021). *TAM: Test Analysis Modules*. R package version 3.7-16.

Shea, T. L., Tennant, A., and Pallant, J. F. (2009). Rasch model analysis of the depression, anxiety and stress scales (DASS). *BMC Psychiatry*, 9(21).

Sijtsma, K. and Molenaar, I. W. (2002). *Introduction to Nonparametric Item Response Modeling*. SAGE, Thousand Oaks. ISBN 9780761908128.

Smith, R. M. (1986). Person fit in the Rasch model. *Educational and Psychological Measurement*, 46(2):359–372.

Smith, R. M. (1996a). A comparison of methods for determining dimensionality in Rasch measurement. *Structural Equation Modeling*, 3(1):25–40.

Smith, R. M. (1996b). A comparison of the Rasch separate calibration and between-fit methods of detecting item bias. *Educational and Psychological Measurement*, 56(3):403–418.

Smith Jr, E. (2020). Detecting and evaluating the impact of multidimensionality using item fit statistics and principal component analysis of residuals. *Journal of Applied Measurement*, 3(2):205–231.

Walker, A. A., Jennings, J. K., and Engelhard, G. (2018). Using person response functions to investigate areas of person misfit related to item characteristics. *Educational Assessment*, 23(1):47–68.

Wilson, M. (2011). Some Notes on the Term: "Wright Map." *Rasch Measurement Transactions*, 25(3), 1331.

Wolfe, E. W. (2013). A bootstrap approach to evaluating person and item fit to the Rasch model. *Journal of Applied Measurement*, 14(1):1–9.

Wolfe, E. W. and Smith Jr, E. V. (2007a). Instrument development tools and activities for measure validation using Rasch models: Part I—Instrument development tools. *Journal of Applied Measurement*, 8(1).

Wolfe, E. W. and Smith Jr, E. V. (2007b). Instrument development tools and activities for measure validation using Rasch models: Part II—Validation activities. *Journal of Applied Measurement*, 8(2).

Wright, B. D. and Douglas, G. A. (1975). *Best Test Design and Self-Tailored Testing.* MESA Psychometric Laboratory. Memo no. 19.

Wright, B. D. and Masters, G. N. (1982). *Rating Scale Analysis.* MESA Press. ISBN 9780941938013.

Wright, B. D. and Mok, M. M. C. (2004). An overview of the family of Rasch measurement models. In Smith, E. V. and Smith, R. M., editors, *Introduction to Rasch Measurement*, pages 1–24. JAM Press.

Wright, B. D. and Stone, M. H. (1979). *Best Test Design.* MESA Press.

Index

Note: Locators in *italics* represent figures and **bold** indicate tables in the text.

Printed in the United States
by Baker & Taylor Publisher Services